Velocette
KSS Mk 2
A Restoration Guide

By Gordon Lund CEng MIET

ISBN 9781783181834

Velocette 1947 factory sales publicity shot

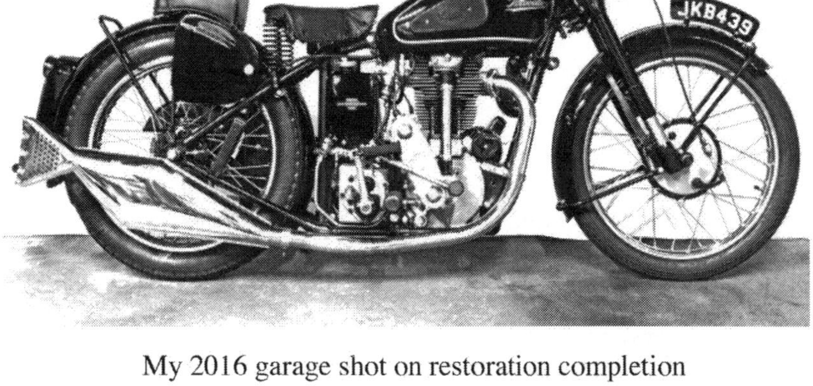

My 2016 garage shot on restoration completion

Foreword

It was with trepidation that I thought of writing this book, as having written a book on my other passion, Lotus Elan cars, my general knowledge on all things two wheels is set in another era. Unlike the Elan Plus 2, which I have owned for many years and the Elan Sprint which I owned for 8 years, restored and maintained by myself out of necessity and general interest. Motorcycles, especially vintage ones had not been a main interest, however an opportunity to purchase a vintage Velocette KSS Mk2 in bits arose and a 5 year transitional journey from a pile of bits to a machine in concours condition began.

As an engineer, part of my early career was spent designing diesel engines for Perkins of Peterborough. That always maintained my interest in engines.

My last major experience on two wheels was in 1969 when I had a Triumph Thunderbird. Before that I had owned a few bikes having passed my motorcycle test at 17 years of age, every one better than the last as my young engineering apprentice's earnings increased, culminating in a BSA Gold Star which nearly ended me never mind my career. Thereafter, my interest in bikes turned towards something a little quirky at that time, which was my first fling with a Velocette, the pre war KSS Mk2. More of that later.

The Triumph came along at the same time as I had my first car, an aging Austin A30. This was superseded with a range of Morris 1000's including a Traveller which was good for carrying prams and everything else that went with early married life. Bikes never crossed my mind for quite some time, until my youngest son, Matthew, had the urge to go motorcycling. Just like myself in my formative years, his friends were also into bikes so that is the path he took. A 400cc Honda VFR just happened to come my way, so that revitalised my interest.

By today's modern bike standards it was still old, on carburettors of all things, but was a little too fast for me. The Honda was OK, but it had nothing that interested me in the tinkering stakes.

Having restored the two Lotus Elans, I thought I would like to restore an old motorcycle but what? I considered Triumphs but even Tiger Cubs were expensive and other bikes like Gold Stars were demanding silly money. The Velocette KSS came back to my mind and I thought to myself now that would be fun. Needless to say that is the way I went.

Having spent 6 years or so restoring the KSS and taking photos at every opportunity, I thought to myself, I wonder if I could write a book on this restoration process. It might interest someone. The bike itself is very much a niche market interest as is the Velocette marque. There are not many KSS machines around so the demand for such a book may be very limited.

I had occasion to talk to my Publisher, John Dowdeswell of Brooklands Books on the subject of my Lotus book when I broached the subject that I had bought a basket case (literally) 1947 KSS OHC Velocette. His first remark was "is there a book in it?". I replied that I had been thinking of it. His retort was "then get on with it!" That was the encouragement I needed so here it is.

First Edition 2018

Published by: Brooklands Books Ltd.

Printed by: Brooklands Books Ltd.

Acknowledgements

To David Matthew Scott Holder of the Velocette Motor Cycle Company Ltd for permission to use the Velocette Logo.

To the Mortons Motorcycle Media, publishers of The Motor Cycle periodical for the use of the Beresford KSS section engine drawing, exploded drawings of the Miller dynamo and BTH magneto and the UBIQUE article on the KTT Mk VIII race engine.

To the Velocette Owners Club for use of extracts, photos and diagrams which have been copied from various web sites and Veloce published Instruction Manuals.

To all members of the Velocette Owners Club who have provided endless encouragement in the KSS restoration and advice given and accepted at every opportunity.

To Ivan Rhodes for confirming the provenance of the bikes registration details from the official works records.

To Colin Goodwyn, VOC Archivist for additional advice on dealing with the DVLA.

To Peter Manley for his proof reading of my draft document and advice on all matters technical, commercial and legal.

To Paul Spare, an engineering colleague who also proof read the draft document.

And finally to my wife Johanna who again lost me to the man cave for yet another restoration project and then providing encouragement and grammatical correctness to my scribblings in producing this book.

2016 garage shot on restoration completion: Courtesy of Matthew Parkinson.

All other photographs, including the Front Cover taken by the author. Where photos and other material sources have not been identified these are not recognised infringement of any existing copyright and if indeed by unintentionally doing so, the author's sincere apologies are offered.

Contents

Introduction

Having rebuilt and partially restored one KSS Velocette and then after a considerable length of time having decided to start another one, I was interested to note that there was very little written about the machine and advice on restoration matters. I had worked on my earlier KSS from nothing more than a little book called the Book on the Velocette by Ferrers Leigh, published by the Pitman's Motor Cyclist's Library and relied heavily on youthful enthusiasm and a smattering of engineering knowledge gained as an engineering apprentice.

The official Velocette publication, the Instruction Book for Velocette Mark II Overhead Camshaft 350cc KSS and KTS, price 2/6d nett, is a useful compendium of information and was written as a guide for then current owners to maintain their machines in tip top condition. It was never intended to be used as a restoration reference many decades ahead.

R W Burgess' book, Velocette Motor Cycles, A Practical Guide Covering Models from 1933, published by C Arthur Pearson Limited is a good practical book which covers all models from the GTP two stroke, LE and Valiant 200cc four stroke twins, the range of push rod singles and the KSS range of overhead camshaft machines (Cammy's) including Mark 1 and Mark II. It does not cover the race variants of the KTT models which would require a complete volume on their own.

The pushrod 350 and 500cc singles are well documented by various authors and an illustrated Haynes Manual gives a good account of all maintenance procedures for these models.

A good pictorial essay of the KSS Velocette is given in Jeff Clew's book Super Profile which quite explicitly gives a good collection of photographs of the KSS range and a succinct essay on the birth of Velocette. The early chapters give a good account of the development of the K series of motorcycles.

To this end, I decided that I could write my own account of the process of the restoration of my own Velocette KSS Mk II. I had taken countless photographs of progress as and when it occurred so please excuse me if the photo quality is not of photo technical journalism standard.

Having gone through the photographic process on the restoration of two Lotus Elan sport cars which turned into a published restoration guide, this book follows the same course.

The Lotus Elan Restoration Guide has been given good reviews and sold well throughout the world. I hope this guide for the much older Velocette KSS, affectionately known to its followers as the Cammy Velo, attracts a similar audience and maybe many more who wish to know what the idiosyncrasies and attractions were for such a well respected make of motorcycle. For such a small manufacturer, Velocette were the giant killers of their day, winning World Championships in the face of much more prestigious names. This was very much like Lotus, decades later in the four wheel sector.

The Author's Motorcycles

My earliest memories of a motorcycle was when I was sat on my uncle's 350cc BSA at the age of four for a photo shoot with a friend of my father behind the lens of his camera.

My motorcycling experience for real started as a youth when I began an engineering apprenticeship on leaving school at the age of 16. Some of the older apprentices had motorcycles but the younger lads had to resort to pushbikes or indeed public transport when the weather got bad in winter. I used to get the odd lift now and again off some of the older lads on their motorcycles, some having only just passed their test and able to carry unqualified riders as pillion. My first ride as pillion was on the back of a 197cc Frances Barnett two stroke. I felt like we were riding the TT course as we buzzed down the country lanes to work.

The author on his uncle's 350cc B31 BSA, age 4

I was bitten by the bug and saved up as best as I could for my first bike. Don't forget that in those days, apprentices were lucky to take home £3 per week in wages, which, had to pay for board (insisted upon by mother) and spending money, etc, left very little for such luxuries as a motor cycle. Some rich kids had their parents buy them one, usually the best they could afford, but the rest of us scrimped and saved for whatever we could afford.

My first bike was a BSA 250cc C11, bought for the sum of £6. It was a non-runner, neither taxed nor insured so I pushed it home for over two miles. I was young and fit then but still goosed when I got it home. Within the space of a week, I had it running and duly taxed and insured it. This started off what can only be called a steep learning curve.

After a few months of riding, a drunken motorist ran into me at traffic lights writing the bike off, breaking my ankle and resulting in my leg in plaster for 6 weeks. This did not deter me so as soon as I could and with the insurance money at hand, I bought a 225cc Francis Barnett Cruiser, complete with Dolphin fairing. Luxury at last. That bike did me good service for a few more months, but, after I passed my test, I wanted something a bit bigger in order to keep up with my friends who were now the owners of BSA Gold Stars and Velocette Venom Clubmans.

Then it was that I signed my life away to purchase a gorgeous 350 cc BSA Gold Star DB32. This I rode and enjoyed for about 12 months when I had an accident which resulted in me being hospitalized for a week. The bike was repaired and duly sold.

My 350cc BSA Gold Star DB32, then girl friend sat side saddle

I took a backward step after that and bought a Lambretta scooter just to get about. I soon got fed up with that and bought an Ariel Leader 250cc twin cylinder two stroke. That lasted a while being a perfectly practical bike but lacked the punch and image of the bigger bikes.

Having a bit more practical knowledge now having served 3 years of my apprenticeship, I decided to build a special. I bought an old clapped out 650 cc Triumph Thunderbird and transplanted the engine into a BSA frame, all at minimum cost. After doing all the sums of buying and selling, it cost me £28 to build it and I sold it for £40. During this time I had passed my car driving test and my father let me use the family car which soon brought home to me the convenience and luxury of not getting wet through and cold in bad weather. Father insisted in time that I bought my own car so I purchased my first car, an Austin A30 for £28. Those were the days.

I still hankered after a motorcycle, still having the bug within me, and so a 250 cc MV Agusta came my way. That was fun as I rode it to work on fine days and spent time tinkering with it to bring it up to what I considered an acceptable standard. At about the same time, a friend at work informed me of the whereabouts of a Cammy Velo. In those days, an overhead camshaft engine was deemed a bit special. The guy had a shed full of interesting old machinery, a bronze head Rudge Ulster and two KSS Velos. One was a runner which he wanted to keep but he would let me have the non-runner for £15. Thus I duly parted with my cash and pushed it home. A bit of déjà vu. That bike taught me a lot about the first principles of motorcycling.

1966 My old "Cammy" Velo complete with "Brooklands" Silencer

My sister Dorothy on the same bike.
Why I turned the handle bars upside down I'll never know

A friend who was emigrating offered me his very nice, restored Triumph T110 650cc Thunderbird for a very good price, so the MV Agusta was sold on to make room. The Triumph was put to good service and ridden for countless miles including a trip to the Isle of Man TT races with my future wife. In this time the Velo was sold on.

I rode the Thunderbird for quite some time but long distance commuting between Preston and Peterborough became increasingly unpleasant as winter approached. My motorcycling days looked over as I traded the Thunderbird in for a Morris 1000 Saloon and I settled down as a happily married man soon to start a family.

A small skirmish with another Lambretta scooter for basic commuting was all the two wheeling I did for a long time. Then out of the blue a few years ago, I was offered a 400cc Honda VFR. As a "born again biker" I was very wary of it. It was a beast compared to everything else I had had before, including the 350 Gold Star. I enjoyed that bike but eventually sold it on to a collector who had every variant of the machine except the one I had. A good deal was struck and it left the fold to make a bit more room for the latest acquisition, my KSS Mk 2 in kit form. I was about to go into hospital for a heart bypass operation so I thought I had better temper my riding ambitions as the years pass by.

My Honda VFR 400 NC21, Johanna my wife astride

That is a brief history of my motorcycling background that gave me the impetus to get down to basics and faithfully restore a classic bike such as the Cammy Velo. Having restored two classic Lotus sportscars, a good classic bike restoration would appear to be a much easier handle-able task. So I thought. The bits might be a lot lighter and take up less room than a car, but as for complexity and awkwardness, if you don't have the patience, don't take on a Velo especially a Cammy. The saying goes something like, "built by engineers for engineers" you have to believe.

Purchase of the Dowty Forked KSS Mk 2

Back in 2010, I advertised in the Velocette Owners Club (VOC) Fishtail magazine for a Mk 2 Velocette KSS. I had one as a young man back in the 1960's that I had bought for the princely sum of £15, had spent £15 on it to get it running and then reluctantly sold it for £45. A Velocette enthusiast had seen the bike parked up outside my parent's house and called in to see if it was for sale. He had heard the engine running and thought this could be worthwhile investigating. Back in the mid 1960's, old bikes were not as popular as they are now and considered a bit quirky. After a bit of negotiation I agreed to let it go. The chap said he only wanted it for the engine.

Looking back now it was one of those if only I..... moments. On reflection I knew that it was something different. The giveaway was the period Dunlop racing tyres, unidirectional, as indicated by the large arrows on the tyre side wall. Also the $1^{3}/_{16}$" Amal TT carburettor, even bigger than the same period KTT race machinery and green racing brake linings. It had no lights or even a dynamo but a pretty good racing magneto.

I had completely restored the engine, parts were affordable then and I had a load of parts rechromed including the wheel rims which I respoked and trued myself. The Webb front fork spindles and bushes were well worn so I made new shafts and bushes to replace the old ones. This transformed the handling of the bike. At the time I was coming up to finishing my engineering apprenticeship as a tool maker and had the full run of the machine shop.

I also had a Triumph T110 Thunderbird in the garage which seemed a far better practical machine so the Velo went on its way.

Back to recent times, I was coming up close to retirement and I was looking for something to take on as a project and considered a classic bike rather than another classic car. I had rebuilt two Lotus Elans and still own the first one, an Elan+2 which I have retained out of sentiment as all my children grew up with that car and it holds lots of memories for me and the family. I had a 400cc Honda V4 at that time but I was not getting any younger, and arthritis in the wrists was not helping with such a low riding position as presented by the Honda. In time, the Honda was passed on and I concentrated my efforts on the Velo.

Back to the advert for a KSS in the Velocette Owners Club Fishtail magazine. I received two replies much to my amazement, one for half a bag of bits including half an engine and the second for basically a complete machine albeit in kit form. I opted to look at the latter, after the owner sent me photos of the bike before it was stripped, dated 14.10.92.! He assured me that it was all there and would I like to see it. As the price was well within my budget I went to see the bike, and took a trailer.

Photo of my KSS, dated the 14th Oct 1992, last fully assembled, given to me by the previous owner.

The vendor was a nice chap who was obviously a classic biker. In his garage was a very tasty 1960's Velo which convinced me that the chap was genuine. Then he took me into his shed where I saw all sorts of things which had me guessing just what I was supposed to be looking at. To one side there was a completely restored Sunbeam S7; and, to the other, a side valve girder forked Norton in need of a major restoration.

Nowhere could be seen the KSS Velo until I looked behind the shed door and there was a KSS engine almost complete bar the magneto and drive casing. Then the vendor showed me the KSS frame and wheels sporting new rims and spokes, with a new tyre on the front.

Engine as found behind the shed door

Frame with rear mud guard and saddles

Then the mudguards appeared, all powder coated in black, along with the frame suffering for not responding well to damp storage over a long period of time. The piece de resistance though was when the vendor opened a chest of drawers in the shed that contained all the other bits that had been stripped off the bike back in 1992. All the tinware was there including the primary chaincases, rear chain guards, oil tank, petrol tank, tool box and the Dowty Oleomatic front forks. Boxes came out of the drawers containing clutch, gearbox (in pieces), magneto, including the chain drive and casing, dynamo, head light with nacelle, saddles, number plates, levers and control cables including an array of nuts, bolts, fixtures and fittings as if the bike had only been recently stripped for rebuild.

Petrol tank, front mud guard and bits in the drawer

Front forks intact sans paint

The vendor then informed me that he had bought it as it was presented to me from the previous owner but had considered it to be a "Bridge to Far" project for him to undertake and had decided to sell it on. The last thing to convince me to take it on, apart from the attractive price being asked, was that the engine appeared in remarkably good condition on close inspection, and the crankshaft turned by hand.

A new registration document with the original buff log book came out with lots of bills and bits of information including a copy of the original Velocette parts manual. The vendor was particular that I would not break the bike up and sell it for parts. I convinced him that I was going to build it so he was happy.

Without further ado, a price was agreed, I'm sure his asking price was reasonable, and the bike was loaded into boxes then into the trailer and transported home. I was now the proud owner of a genuine 1947 Velocette KSS Mk2.

Having joined the Velocette Owners Club prior to this purchase, I registered the bike with them and requested verification of the registration number, engine and frame number. As expected, I received from the VOC a letter confirming the provenance of the bike signed by the Club President, Ivan Rhodes.

Everything in the garage at home

A Brief History of the Velocette Models

Model	Engine cc	Year Introduced	Comments
2 H.P. Veloce	200	1907	First motorcycle produced by Veloce
Model D1	220	1919	First two-stroke motorcycle produced by Velocette
KSS Mk 1	348	1925	4 Stroke OHC single iron head
KSS Mk 2	348	1935	4 Stroke OHC single aluminium head
KTT	348	1925	TT version of the KSS in many marks up to the MkVIII
GTP	249	1930	2-stroke engine with coil ignition
MOV	248	1933	High camshaft 4-stroke
MAC	349	1933	OHV air-cooled single
MSS	495	1935	OHV air-cooled single
LE	149	1951	Water cooled horizontal opposed twin cylinders
Venom	499	1956	Sports model
Viper	349	1956	Smaller capacity version of the Venom
Valiant	200	1956	OHV LE engine air cooled in tubular frame
Viceroy	249	1960	2 stroke scooter
Vogue	192	1963	LE with GRP bodywork
Thruxton	499	1965	Velocettes swansong. The ultimate Venom

The Development of the K Series

The K Series Velocette motorcycle engine was conceived by Percy Goodman, son of Velocette founder Eugene Goodman, as a wish to go further up market from the two stroke engines they had developed up till that time. The engine was deemed a success right from the start and proved to have considerable potential proved by early TT successes in the 1920's.

The factory race engines were designated KTT to differentiate them from the KSS models or Super Sport. As the KTT was developed at a pace they left the KSS behind in the performance stakes as one would expect from the priorities in race workshops.

The KSS carried on as a sports tourer and was refined into the Mk 2 with improvements such as an aluminium cylinder head, enclosed valve gear, stronger gear box and drive train. An interim mark of the KTT actually utilized the MK 2 KSS cylinder head until the factory race shop produced their own cylinder head for the KTT.

Due to the race successes of the factory race team, replicas were produced to meet the demand of the amateur racers which proved to be a commercial success for Velocette. At that time they were the only manufacturer to offer for sale to the general public what were ostensibly copies of the factory race machinery. They were so popular that, in some races including the Isle of Man TT, they filled a large proportion of the starting grid and hence a large proportion of podium finishers. The history books record Velocette's dominance of the 350cc World Championship by winning a number, particularly between the war years.

The 2nd World War was not kind to Velocette as they were not really in a position to produce the extremely large numbers of machines required by the armed forces, unlike BSA and Triumph and, to some extent, Norton and other smaller manufacturers. As such, after the war, Velocette soldiered on as best they could with outdated manufacturing facilities and limited financial reserves. They had some limited success in the late 40s, early 50s with the KTT Mk 8, but could not keep pace with Norton and the up and coming Italian marques. Their challenge in the 500cc class with the twin cylinder supercharged "Roarer" fell on stony ground as superchargers were banned after the war. Velocette then gave up the factory sponsored race team and concentrated on their mainstream models.

Attempts to capture the imagination of the motorcycle buying public with derivatives of the 200cc LE proved to be a financial disaster and placed the final nail in the coffin as Velocette, along with many other British manufactures, succumbed to the invasion of Japanese machinery.

Even the evolution of the 500cc Venom Thruxton into the very successful production race machine that emanated the Gold Star discontinued by BSA in the early sixties was too little too late: and, by the mid 60's, the writing was on the wall for Velocette.

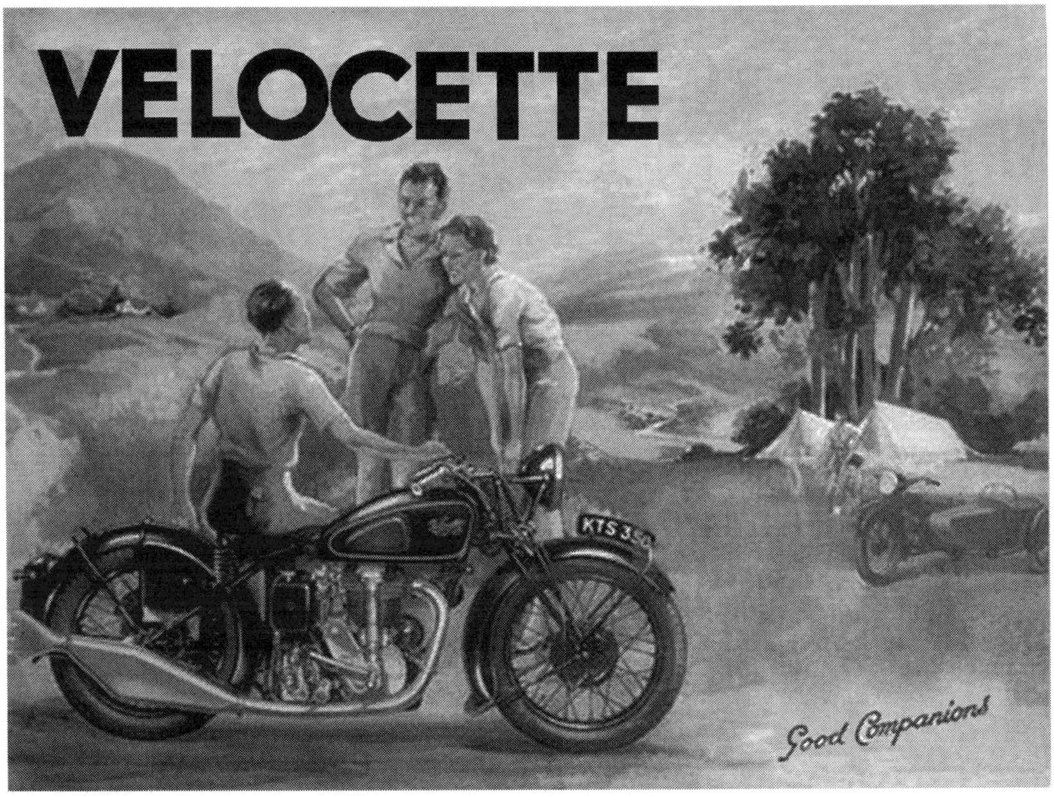

Pre war KSS Mk 2 advertisement

Note the hair pin valve springs & the ribbed crankcase

An early KTT in period race trim

Pre war KSS Mk2 on the road

An early KTT *Note ribs on the crankcase*

1947 Velocette KTT Mk VIII

1950's Velocette "Roarer"

Velocette "Venom"

Velocette "Venom Thruxton"

Velocette 1931 Factory Brochure

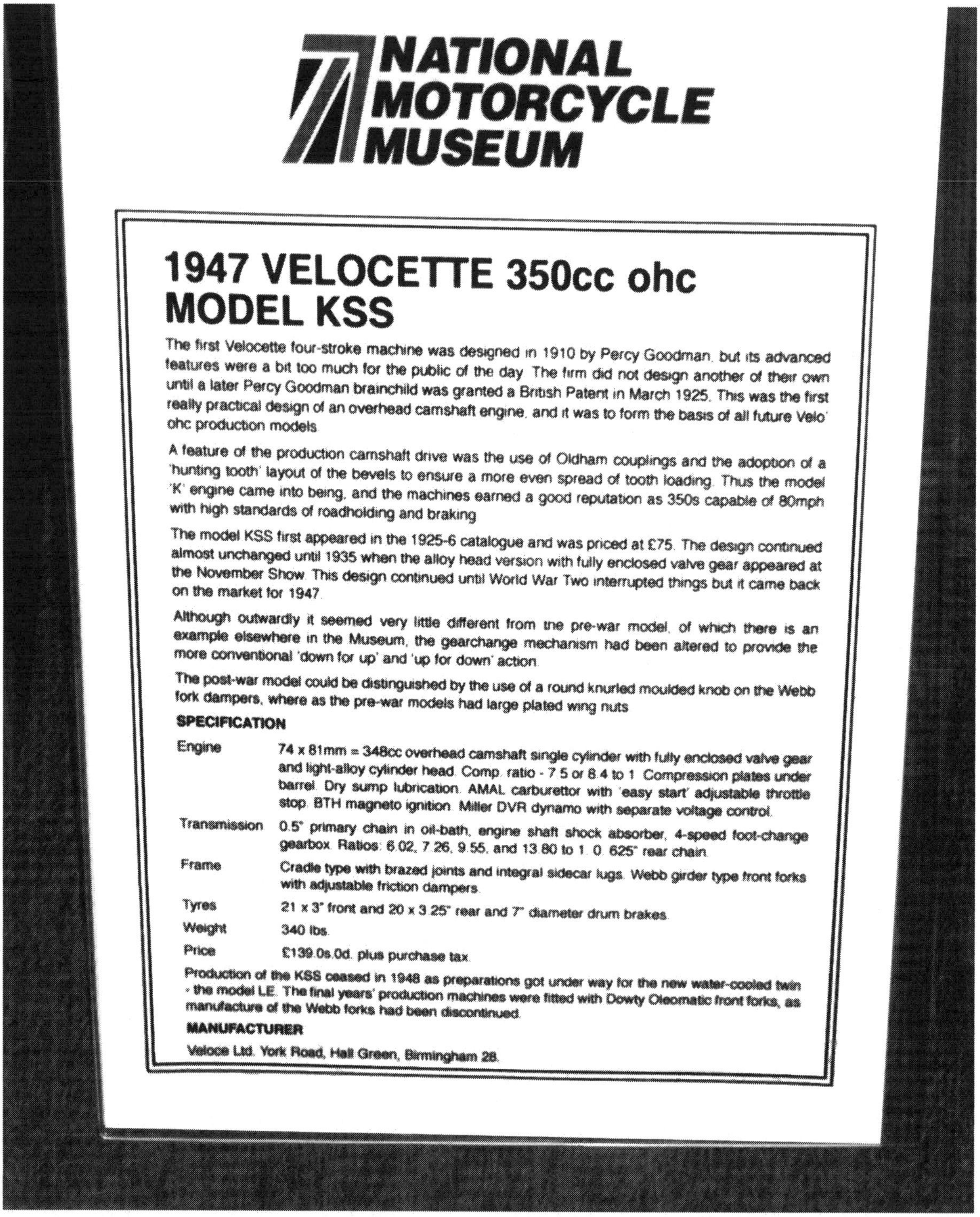

NATIONAL MOTORCYCLE MUSEUM

1947 VELOCETTE 350cc ohc MODEL KSS

The first Velocette four-stroke machine was designed in 1910 by Percy Goodman, but its advanced features were a bit too much for the public of the day. The firm did not design another of their own until a later Percy Goodman brainchild was granted a British Patent in March 1925. This was the first really practical design of an overhead camshaft engine, and it was to form the basis of all future Velo' ohc production models.

A feature of the production camshaft drive was the use of Oldham couplings and the adoption of a 'hunting tooth' layout of the bevels to ensure a more even spread of tooth loading. Thus the model 'K' engine came into being, and the machines earned a good reputation as 350s capable of 80mph with high standards of roadholding and braking.

The model KSS first appeared in the 1925-6 catalogue and was priced at £75. The design continued almost unchanged until 1935 when the alloy head version with fully enclosed valve gear appeared at the November Show. This design continued until World War Two interrupted things but it came back on the market for 1947.

Although outwardly it seemed very little different from the pre-war model, of which there is an example elsewhere in the Museum, the gearchange mechanism had been altered to provide the more conventional 'down for up' and 'up for down' action.

The post-war model could be distinguished by the use of a round knurled moulded knob on the Webb fork dampers, where as the pre-war models had large plated wing nuts.

SPECIFICATION

Engine	74 x 81mm = 348cc overhead camshaft single cylinder with fully enclosed valve gear and light-alloy cylinder head. Comp. ratio - 7.5 or 8.4 to 1. Compression plates under barrel. Dry sump lubrication. AMAL carburettor with 'easy start' adjustable throttle stop. BTH magneto ignition. Miller DVR dynamo with separate voltage control.
Transmission	0.5" primary chain in oil-bath, engine shaft shock absorber, 4-speed foot-change gearbox. Ratios: 6.02, 7.26, 9.55, and 13.80 to 1. 0.625" rear chain.
Frame	Cradle type with brazed joints and integral sidecar lugs. Webb girder type front forks with adjustable friction dampers.
Tyres	21 x 3" front and 20 x 3.25" rear and 7" diameter drum brakes.
Weight	340 lbs.
Price	£139.0s.0d. plus purchase tax.

Production of the KSS ceased in 1948 as preparations got under way for the new water-cooled twin - the model LE. The final years' production machines were fitted with Dowty Oleomatic front forks, as manufacture of the Webb forks had been discontinued.

MANUFACTURER

Veloce Ltd. York Road, Hall Green, Birmingham 28.

NMM Display Description of a KSS exhibit

Restoration of the Author's KSS

Frame and Tinware

The rear mudguard was assembled onto the frame when I took possession of the bike. On careful inspection it became obvious that the powder coating undertaken by one of the previous owners was suspect with pockets of rust showing through. It was imperative that this be rectified before any progress was to be made in the rolling chassis restoration. Subsequently, the rear mudguard was removed and put to one side.

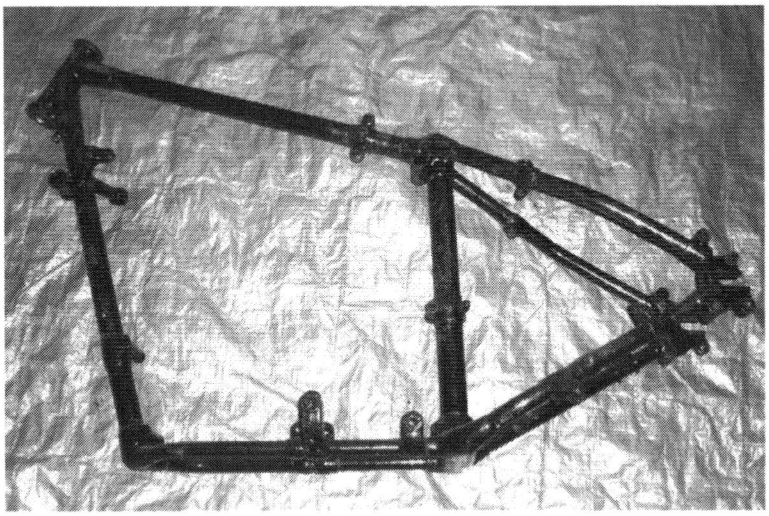

Suspect powder coated frame

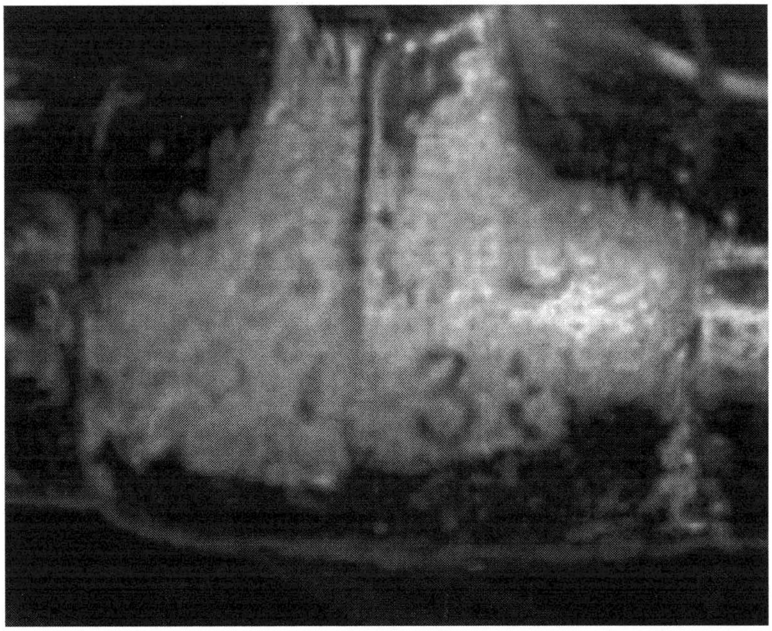

Frame number identified

Powder coating removed

All of the powder coating was removed with paint remover, scraped and washed off with water and finally wire brushed to remove all remnants of coating and rust. The frame was then subjected to plenty of rubbing down with emery cloth to smooth off the surfaces as much as possible to allow close inspection for signs of cracking and any deficiencies in the brazing at the lug joints. It also allowed me to uncover and verify the frame number. The frame was then sent for powder coating in the obligatory black.

New bright black powder coating

The mudguards, stays, chainguards, headlight nacelle and other tinware was subjected to the same treatment prior to either re powder coating or paint spraying. It must be said at this point that I tried powder coating for most of the steel items but found in some cases that this did not give the degree of finish I required that befits a machine of this quality, especially for show purposes. Therefore some items such as the headlight nacelle and other minor items have been subject to both powder coating and spray painting. In addition a shiny gloss coating was achieved using a two pack resin paint to suit the petrol and oil tanks.

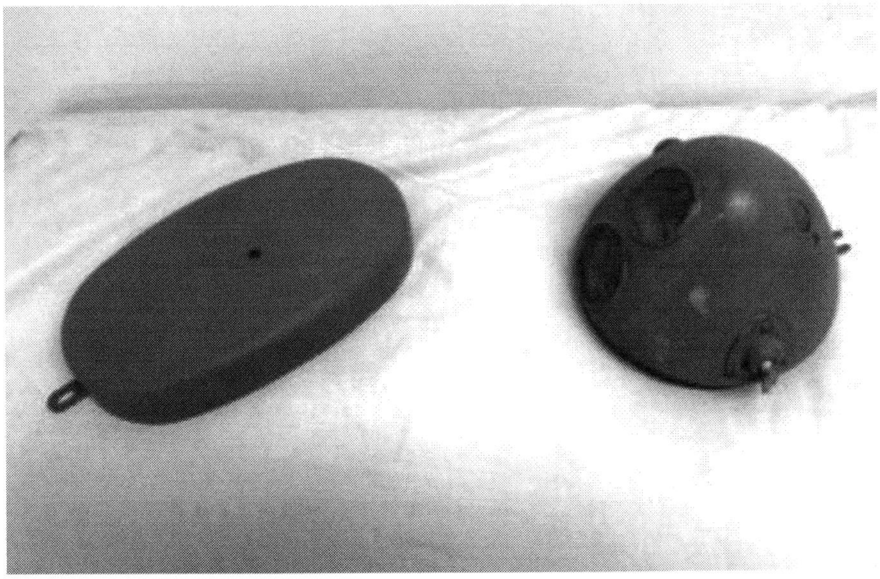

Dynamo belt cover & headlight nacelle in primer paint

Dynamo belt cover finished in high gloss black and clear lacquer

Headlight nacelle finished in high gloss black and clear lacquer

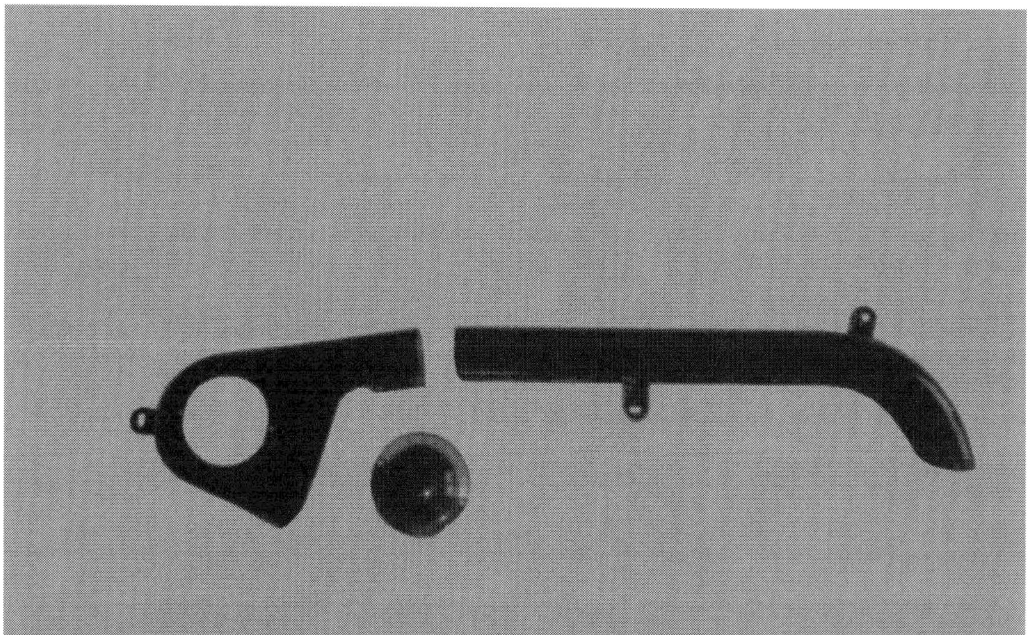

Rear chain guard components finished in high gloss black and clear lacquer

An item that really surprised me as to its original condition was the primary chaincase. The 2BA tapped holes in the rear half were all in perfect condition as were the screw heads on the screws I found in a tin with the rest of the bits and pieces. Bar for the poor condition of the plating that had disappeared off the screws over time, I would have reused them. The front cover was in remarkably good condition as well apart from the rusty paintwork.

Primary Chaincase outer and inner prepped for painting

I put this down to the relatively small mileage that the bike had done, 38,000 miles indicated on the clock. Whether or not that is the true mileage is open to conjecture. The cylinder bore was 0.030" oversize, but that does not give a complete picture given the age of the bike. It was standard practice that de-cokes were done on a regular basis in those days, 0.003" oversize would have been acceptable in what would be considered a small mileage by today's standards. Given the roads of the period, the bike looks to have had little use and was probably well maintained in its early years. Most of the deterioration evident when I purchased the bike was probably due to poor long term storage conditions rather than abusive use. That would explain the absence of dents and scrapes other than the grounding marks on the exhaust and silencer.

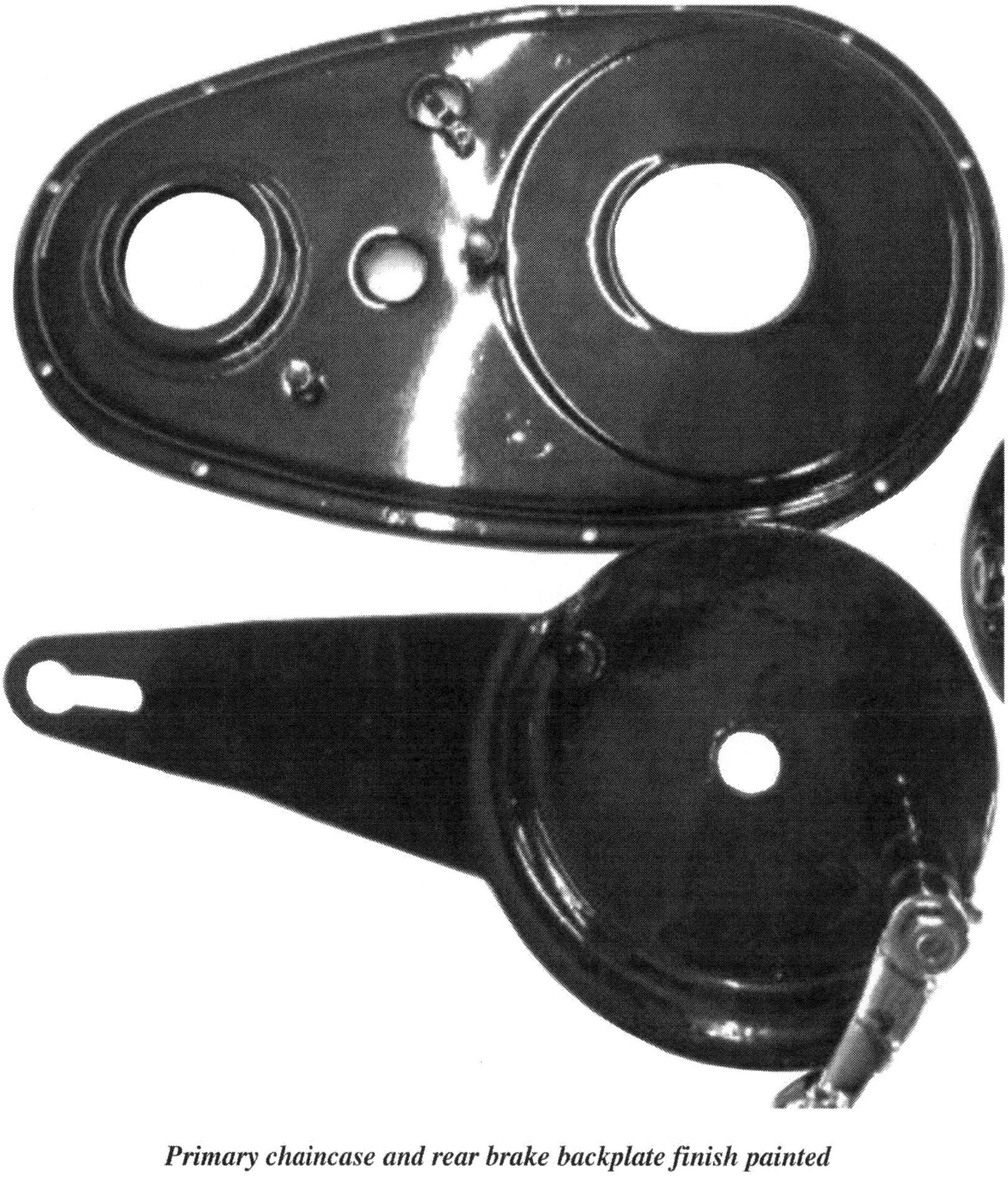

Primary chaincase and rear brake backplate finish painted

Webb Girder Forks

Webb Girder forks were de-rigueur on all pre war Velocettes. My first KSS had them fitted though they were well worn and required some work. I made new spindles out of ground stock EN8 Silver Steel on which I turned down and threaded the ends with BS Cycle threads as per the originals. I also made replacement phosphor bronze bushes. The bike handled a lot better after. The essence of looking after girder forks is to keep them well greased and adjusted and never have the fork damper too tight, as this will affect the handling. Let the forks do the work they were designed to do.

Webb's SUPER SHOCK ABSORBER SPRING FORKS.

DE LUXE MODEL SPARE PARTS.

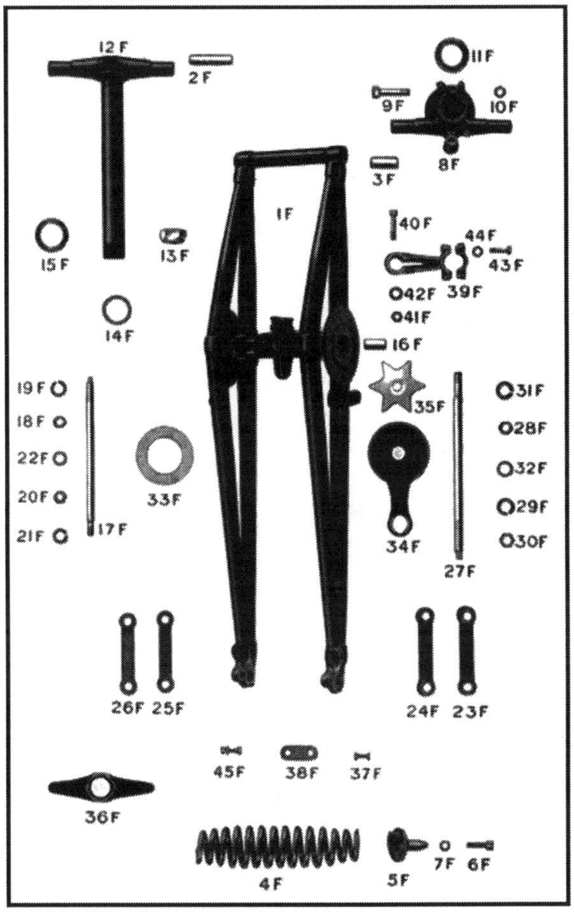

Webb's SUPER SHOCK ABSORBER SPRING FORKS.

DE LUXE MODEL.

List of Spare Parts.

1.F. Fork Girder.
2.F. Steering Column Bearing Bush.
3.F. Girder Bush-top.
4.F. Fork Spring.
5.F. Spring Lug. Top.
6.F. Spring Lug Pin.
7.F. Spring Lug Pin Washer.
8.F. Head Clip.
9.F. Head Clip Pin.
10.F. Head Clip Pin Nut.
11.F. Head Clip Ball Race.
12.F. Steering Column.
13.F. Steering Column Head Nut.
14.F. Steering Column Washer.
15.F. Steering Column Ball Race.
16.F. Girder Bush-bottom.
17.F. Top Spindle.
18.F. Top Spindle Nut. R/H Side.
19.F. Top Spindle Spring Washer. R/H Side.
20.F. Top Spindle Nut. L/H Side.
21.F. Top Spindle Spring Washer. L/H Side.
22.F. Top Spindle Plain Washer.
23.F. Bottom Link. R/H Side. (Screwed).
24.F. Bottom Link. L/H Side.
25.F. Top Link. R/H (Screwed).
26.F. Top Link. L/H (Plain.)
27.F. Bottom Spindle.
28.F. Bottom Spindle Nut. R/H Side.
29.F. Bottom Spindle Spring Washer. R/H Side.
30.F. Bottom Spindle Nut. L/H Side.
31.F. Bottom Spindle Spring Washer. L/H Side.
32.F. Bottom Spindle Washer. Plain.
33.F. Friction Washer for Damper.
34.F. Pressure Plate.
35.F. Pressure Plate Star Spring.
36.F. Hand Adjuster to Damper.
37.F. Mudguard Fixing Pin and Nut.
38.F. Mudguard Fixing Pad Plate.
39.F. Adjustable Handlebar Lugs.
40.F. Adjustable Handlebar Lug Fixing Bolt.
41.F. Adjustable Handlebar Lug Fixing Bolt Nut.
42.F. Adjustable Handlebar Lug Fixing Bolt Washer.
43.F. Handlebar Fixing Bolt to Adjustable Lug.
44.F. Handlebar Fixing Bolt to Adjustable Washer.
45.F. Anchor Pin. Nuts and Washer.

A Typical Webb Girder Fork Arrangement Parts List

The Dowty Oleomatic Forks

Dowty forks were introduced in 1947 as a stop gap before Velocette were in a position to introduce their own sprung front forks. Webb at about this time had stopped supplying girder forks to Velocette who approached Dowty to supply their telescopic forks for the current range of single cylinder machines, including the KSS. Dowty had close links with Velocette providing both rear and front suspension units for the race programme.

The forks on my machine were fully assembled be it devoid of any paintwork when I took the bike home. I cleaned and painted them having established that they appeared to function correctly after I had filled them with the correct amount of oil, and pressurised them with air from a foot pump. I assembled them to the then finished painted frame with new steering head bearing balls and races and thought that would do. Needless to say they did not hold the air for very long and thus required pumping up on a regular basis.

Forks partially stripped

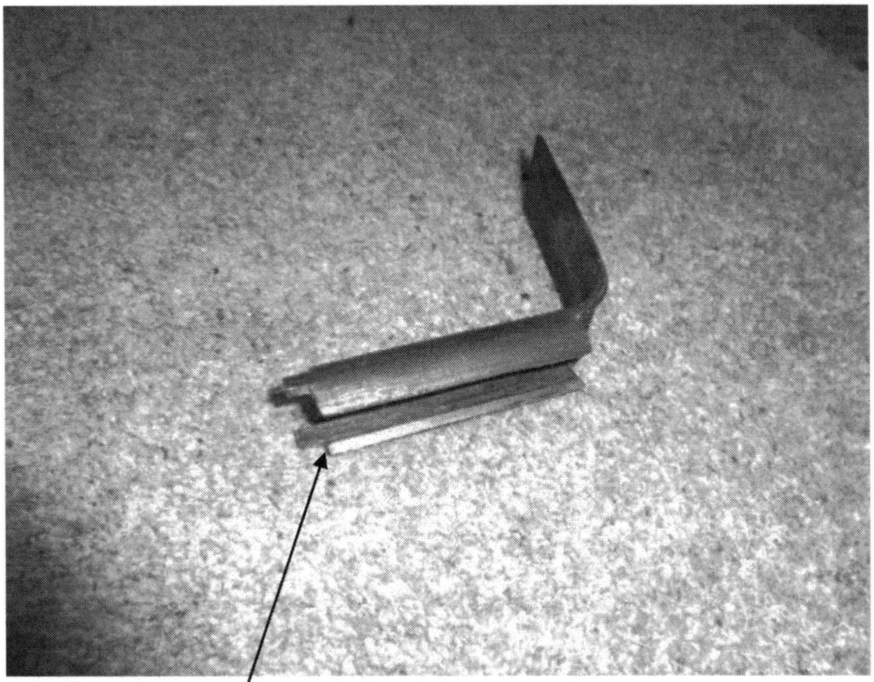

Special tool I made out of steel tube

Seals and damper component parts

A new set of seals was acquired then I set to stripping the forks into their component parts. I made a special tool to engage with the slots in the seal carriers that are screwed into the lower fork stanchions. Having taken all the internals to pieces I inspected the upper stanchion air feed cross pipe only to find a small fracture in the copper pipe at the silver soldered joint into the stanchion fitting. That explained the quick loss of air springing. I made up a new piece of pipe, took out the old broken one with an acetylene torch and silver soldered the new pipe in position.

Stripped forks ready for inspection

Now that the forks were completely stripped, they were totally cleaned inside and out and all non-bearing parts painted with a fresh coat of two pack paint and lacquer. The forks were then reassembled with new seals, filled with oil as per the book and pumped up to the red dot on the lower fork stanchion. The chrome on the lower fork stanchions was in perfect condition, again, another indication of the bikes low mileage. The chrome must have been good to begin with to be unaffected by the damp storage conditions that the bike was subjected to.

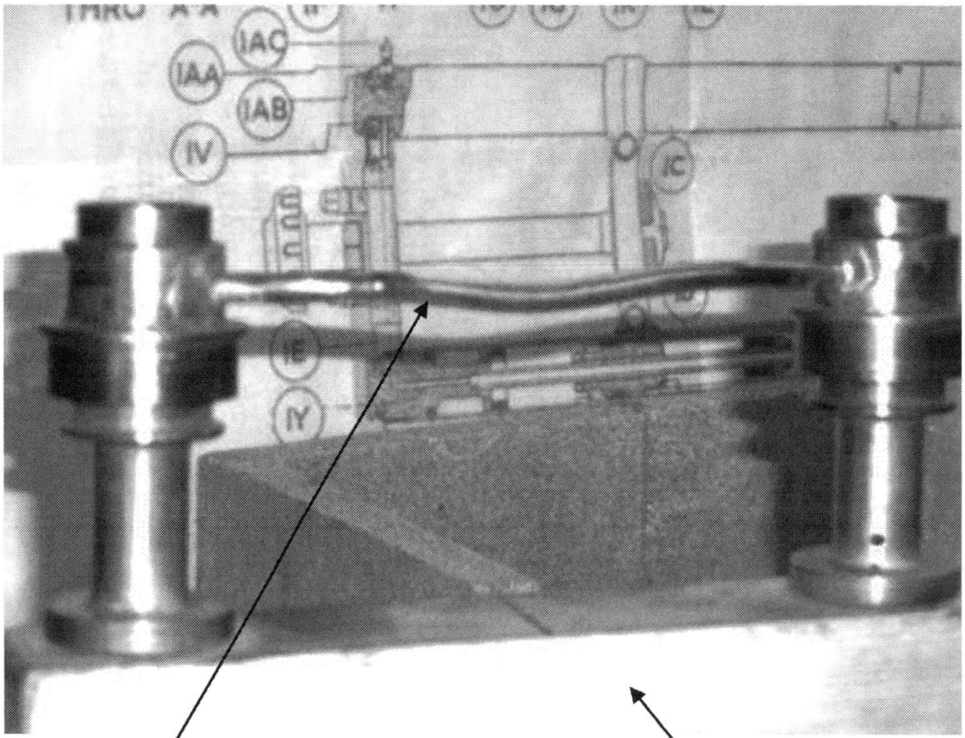

New balance pipe silver soldered in place (note wooden jig)

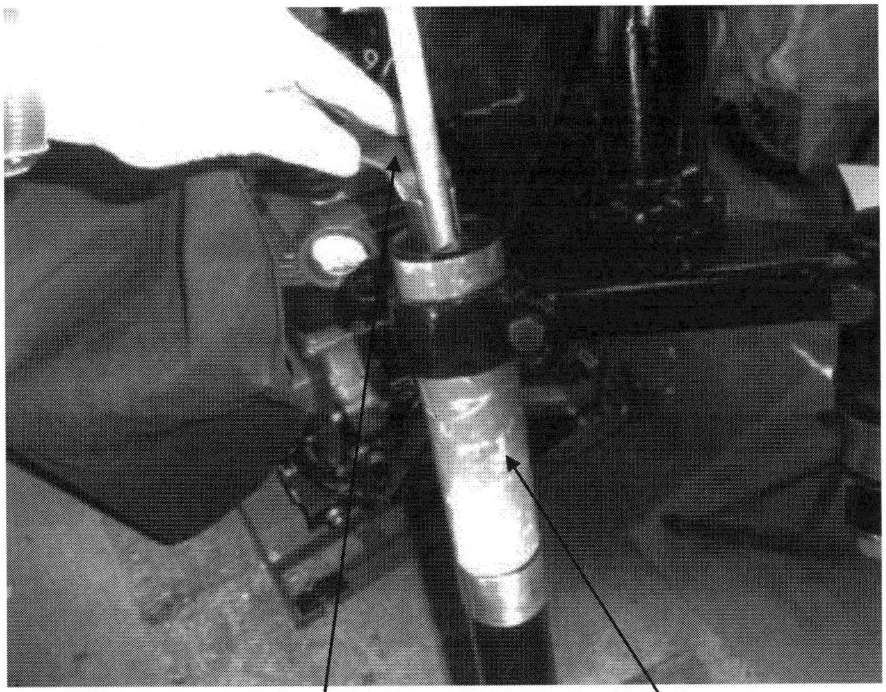

Forks reassembled with special tool (note greased paper to protect paint)

The bike's forks are now working correctly and holding air pressure. These forks have come under severe criticism over the years for losing the compressed air quickly. I would put that down to poor maintenance and nothing else. They do not respond favourably to neglect.

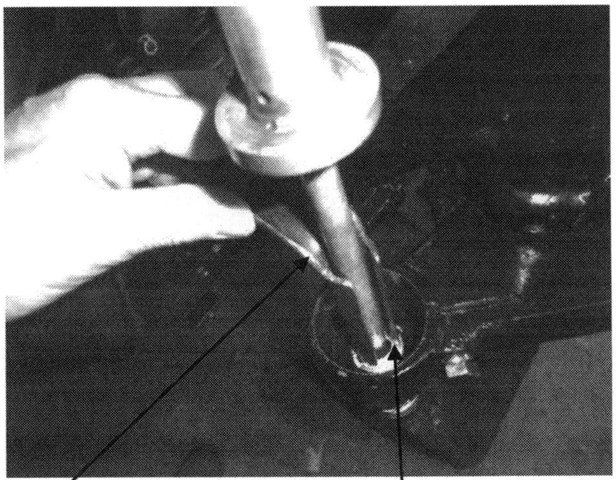

Special tool engaged in slots in screwed seal carrier.

Finished forks pumped up

Dowty made undercarriage suspension legs for WW2 military aircraft. Some of these aircraft crashed in the Pennines during the war; and, apparently, some remains can be seen today with the undercarriage suspension chrome plating still in good condition. Chrome plating back then was not just for decorative purposes as is the case today, but made a tough hard wearing surface.

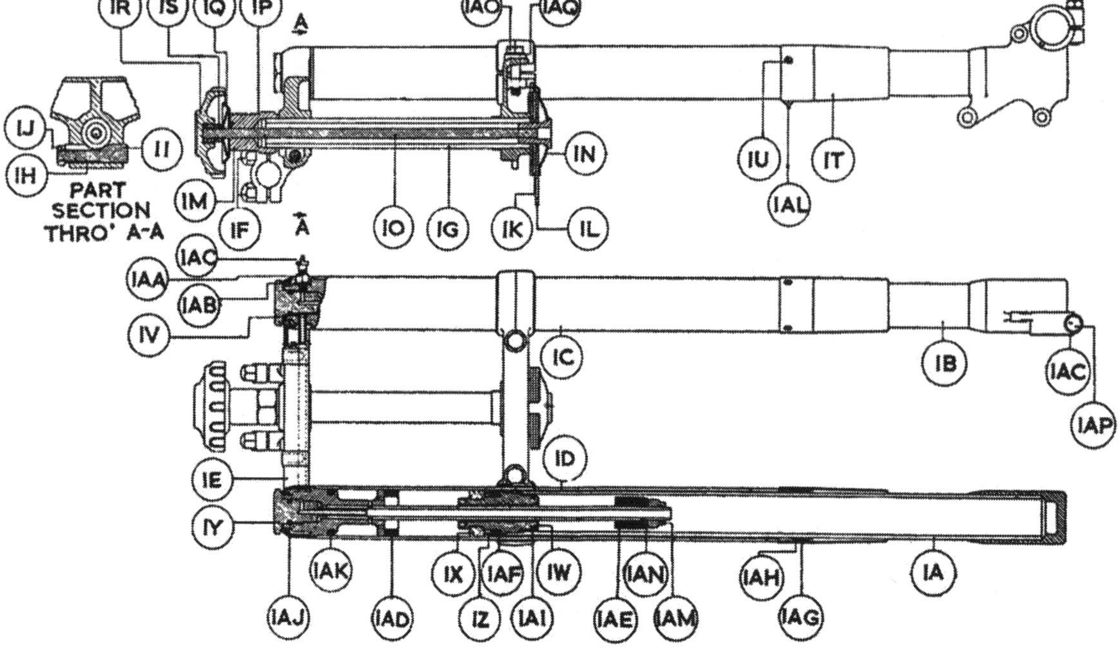

Assembly drawing of Dowty Forks

IA. Axle fitting and sliding member (R–H).
IB. Axle fitting and sliding member (L–H).
IC. Outer tube (L–H).
ID. Outer tube (R–H).
IE. Top-clip lug.
IF. Handlebar bolt.
IG. Steering column.
IH. Clamping-bolt sleeve.
II. Clamping bolt.
IJ. Clamping-bolt nut.
IK. Steering-damper friction disc.
IL. Steering-damper friction plate.
IM. Damper locating plate.
N. Bottom nut and plate for steering damper.
IO. Damper rod.
IP. Steering-column nut.
IQ Damper star spring.
IR. Damper knob.
IS. Damper knob lock-nut.
IT. Shroud.
IU. Shroud screw.
IV. Balance pipe and top fitting units.
IW. Piston.
IX. Upper bedring.
IY. Filler plug.
IZ. Gland ring spacer.
IAA. Inflation-valve body.
IAB. Inflation-valve seal.
IAC. Inflation-valve cap.
IAD. Dashpot buffer.
IAE. Lower buffer.
IAF. Gland ring.
IAG. Scraper ring.
IAH. Scraper-ring packing seal.
IAI. Piston static seal.
IAJ. Filler-plug static seal.
IAK. Top fitting static seal.
IAL. Greaser.
IAM. Lower buffer nut.
IAN. Lower buffer nut washer.
IAO. Column outer tube bolt.
IAP. Spindle clamping bolt.
IAQ. Spring washer.

Legend for above drawing of Dowty Forks

Wheels

When I took delivery of the bike, it became apparent that some early restoration had begun on the wheels at the same time as the frame and some of the tinware. The powder coating was suspect on the frame so that was redone as outlined earlier. The wheels had been completely renovated by one of the previous owners, stripped, cleaned with new bearings inserted in the hubs to replace what must have been well worn and possibly corroded ball races. New chromed rims had been purchased as evident by invoices that came with the bike and details of wheel builders. The hubs had been rebuilt along with the new rims and new spokes. Luckily these had been coated with grease which had protected the chrome and finishes of the spokes, rims and hubs.

The front wheel came complete with a new ribbed front tyre which did not show any signs of aging so must have been stored well out of daylight and the damaging sun rays that cause cracking of tyre walls over long periods of time.

Having closely inspected the wheels for any obvious signs of damage and out of true both laterally and radially, I removed, cleaned, and inspected the wheel bearings, for signs of any wear and corrosion. They had been well packed in bearing grease so had retained their new appearance, confirmed by the finish in the races and no pitting evident on the balls or races when viewed under a magnifying glass. The bearings were then repacked with bearing grease and reinserted into the hubs. Reference to the Velocette instruction manual and the Munroe Book on the Velocette confirmed that the spacers and grease retainers were all present and correct and assembled in the correct order.

The front wheel integral brake drum was inspected and deemed serviceable so was cleaned up and assembled with the renovated and relined brake plate assembly (see later chapter).

The rear wheel was given the same treatment as the front, the brake drum being a separate item. (see later chapter). The absence of a rear tyre meant searching for a suitable one to complement the front tyre and the period of the bike. The Vintage Tyre Co. came up with the goods in a 20 x 3.25 inch Avon with a period tread pattern for this type of machine. It was not cheap, being of a higher speed rating than required, in fact noted as being recommended for Manx Nortons of the same vintage.

Engine

The overhead cam engine is a jewel of engine design and engineering. As such, work on it can be fraught with all kinds of challenges and unexpected events for the uninitiated and non-technical restorer. It is not merely a question of where to start the engine strip but how and why and the set up procedures that must be followed if disaster and costly mistakes are to be avoided in both time and money.

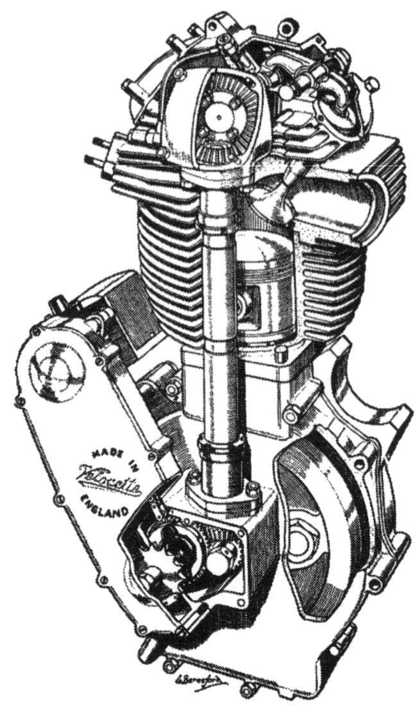

BY KIND PERMISSION OF MOTOR CYCLE
(The Piston illustrated is of a type now superseded by a "Slipper" Piston).

The same dismantling procedure is to be followed for a complete engine strip and a routine decoke. This is because of the peculiarity of the cam drive line. Being a vertical bevel drive, it would have been simpler for Velocette to have made the bevel gears top and bottom on the vertical shaft with the same number of teeth, but they decided to do it right by introducing the "hunting tooth" being an extra tooth in the bevel drive to distribute the valve spring loadings periodically around the vertical shaft plain bearings. It takes 23 turns of the crankshaft to bring the timing marks of the cam shaft drive bevel gear back into line. Because of this, the engine suffers very little wear in this region. The oiling system also helps by feeding oil continuously at engine pressure through slots in the vertical shaft bearings, then onto the cams, and rockers intermittently through a slot in the camshaft plain bearing and a flat on the camshaft journal.

Prior to commencing any strip down, ensure that the engine is cleaned of all debris, oil and general road filth. A good proprietary engine cleaner and a stiff brush will suffice. This will help in keeping unnecessary detritus out of the engine.

Due to this engine design feature, it is absolutely necessary to "time the engine" so that the timing marks on the top bevels coincide with each other while the crankshaft is set at top dead centre. This should also coincide with the slot for the top Oldham coupling being in line with the engine crankcase. This procedure is essential even for a head off only job, to ensure that the timing is not disturbed or incorrectly set at reassembly with disastrous results. The crank must not be turned after the timing gears have been set.

Cylinder head removal is achieved by undoing four ¼" BSF nuts from the studs that retain the top bevel cover, and carefully removing the cover so as not to damage the gasket faces on the head and cover. Next, with the aid of the special two pronged "C" spanner, available from Velocette specialist and the owners club spares department (do not use hammer and chisel / screwdriver!), undo both sleeve nuts; one at the bottom, and one at the top of the cam drive cover tube. Push the top sleeve nut down and pull the bottom one up to expose the spring wire clip in the groove at the bottom of the tube. With the aid of a small screwdriver, release the spring clip and push upwards. This will allow the sleeve tube to be pushed down into the bottom bevel housing, thereby making the top Oldham coupling clearly visible.

Engine in stand with top bevel drive cover held on with four nuts

Nuts removed with cover exposing camshaft bevel gear

Timing marks on top bevels with engine set at top dead centre

Head removed on the bench with sleeve nut C spanner to the right

One could argue that setting the valve timing is not absolutely essential if the engine is being stripped down completely, but it is comforting to know that the engine was timed correctly prior to the stripdown by some other previous owner. Even for record purposes it gives confidence that the engine can be easily reassembled with the minimum of fuss. There is nothing more frustrating as finding a combination of extra timing marks on the bevel gears put there years ago by an inexperienced engine builder.

Having set the engine up ready to remove the cylinder head and all ancillaries such as carburettor, exhaust, petrol tank, engine steady bracket from cylinder head to the frame top tube and control cables, slacken the four cylinder head screws evenly in a diagonal pattern. Use the special tube spanner as supplied by Velocette or a 3/8" drive socket with extension. Do not use anything else or you will risk breaking cooling fins on the head. The cylinder head can now be removed from the cylinder barrel. This operation can be done with the engine in the frame as for a decoke or to make the engine lighter for removal purposes. My preference is for removal on the workbench where there is a better working environment in cleaner conditions. Place the cylinder head in a safe area for later attention. Note that many KSS cylinder heads have broken cooling fins as a result of careless handling and storage.

Drain the oil tank by removing the small drain plug at the rear of the oil tank. A funnel and receptacle held in place is useful as this can take some time as the drain hole is quite small. If the engine has just been run the hot oil will drain better. If the engine has a tendency to wet sump, draining the tank at this time is preferable as there will be less oil left in the crankcase. Disconnect and remove the oil feed and return pipes from the engine from / to the oil tank, then remove the breather pipe from the base at the rear of the tank that terminates over the rear chain. Note that this breather pipe is held onto the rear guard valance by a P clip.

Disconnect the wiring loom from the dynamo. Label the cables to assist with reassembly.

Remove the dynamo drive belt outer guard.

Remove the dynamo and drive belt by undoing the bolts holding the dynamo clamp bracket to the engine having first disconnected the voltage control unit wiring. Label the cables to assist with reassembly

Primary drive side with dynamo removed

Take out the split pin from the crankshaft ring nut that holds the dynamo pulley. Select top gear and step on the rear brake to prevent the crankshaft from rotating. Undo the crankshaft sleeve nut with the aid of the special Velocette four tongue wrench, and remove the dynamo drive pulley, "cush" drive spring, "cush" drive shock absorber outer and inner pieces from the crankshaft.

Remove the battery and the battery platform with battery rear strap from the rear engine plates. This enables removal of the engine plates and improves access to the primary chain case fixings.

Remove the rear chain guard front and rear sections and remove the split pin and castellated nut from the rear chain gear box sprocket. This is best done by stepping on the rear brake to prevent the sprocket from rotating.

Engine & gearbox on the bench

Normally the engine and gear box are removed together having first removed the primary drive, clutch and chaincase (see chapter on Primary Drive for details). Position a jack to take the weight of the engine and gearbox and remove the front engine plates. Undo all the nuts holding the rear engine plates to the frame and remove engine and gearbox as one unit. Assistance may be required as the assembly is quite heavy and cumbersome, especially with the cylinder head in place. Place on the bench, preferably in a purpose built stand for maximum stability, (see photograph of my home made engine stand).

Remove the rear plates fixing the engine to the gearbox. Put the gearbox to one side for attention later. Finish fixing the engine to the engine stand at the rear.

Tightening the timing sprocket nut with the aid of an aluminium "sprag" (put the "sprag" on the other side of the sprocket to undo)

Start the engine strip by removing the screws holding the magneto drive cover to the inner casing. Check the condition of the magneto drive chain and sprockets checking the sideways movement of the chain to assess free play excess. Sprag the magneto drive chain sprockets with aluminium bar of suitable size to prevent the sprockets from rotating and preventing placing excess torque on the timing drive train when undoing the sprocket nuts. (see photographs)

The Velocette Instruction book makes reference to using two screw drivers to remove the magneto sprocket. I have drilled and tapped two holes in the sprockets to enable the use of a two bolt puller to make this job easier when retiming of the engine in the future. (see photograph). For the purist, these holes might also serve to locate a peg socket for future removal of the timing chain / sprockets without spragging them. The lower sprocket is on a keyed tapered shaft. The two sprockets and endless chain are removed as one.

Undo the magneto fixings to the drive casing and remove the magneto to a safe and dry storage area.

Two bolt puller in position on magneto sprocket

Special puller on magneto sprocket

Lower magneto drive sprocket

Undo and remove the nut and washers from the magneto drive casing steady which is attached to the crankcase. Undo and remove the screws holding the magneto drive casing to the crankcase and then carefully remove the inner drive casing. The magneto drive shaft with gear will come away with it along with the Oldham drive coupling contained within the oil pump drive gear.

Magneto drive back plate

Keep all these bits and pieces in lidded tins suitably labeled then you can find the bits afterwards. It keeps dirt out and helps prevent pieces getting lost. Keep the screws etc with the sub assemblies as you strip them, partially assembled then you know where the screws and fixings belong. Clean out the timing chest and look out for extraneous bits of metal and other tell tale debris which should tell you if the engine has had a hard life. If the engine internals are reasonably clean you have been lucky.

Remove the lower bevel housing carefully by releasing the two ¼" BSF nuts from the studs and tapping the housing with a hide mallet.

Cleaned out lower bevel drive with oil pump drive coupling

Next, remove the blind nut holding the spur and bevel gears onto the crankshaft. Note this is a left hand thread, i.e. rotate clockwise to undo. There is a locking tab washer under this nut that will have to be straightened before the nut can be released. Undo and remove the nut carefully. Inside the nut is a ¼" dia ball bearing held against a tapered seat inside the nut with a coil spring inside the internal drilling in the crankshaft. This provides oil to the big end bearing via a hole in the side of the nut whilst ensuring oil pressure is maintained in the timing case for camshaft lubrication.

Blind nut holding oil ball valve and spring to crankshaft

Remove the spur gear with the aid of a puller if required followed by the bottom bevel gear. Note there are shim washers behind the bevel gear that should be retained for reassembly purposes. Retain the key that is located in the crankshaft that slots in the keyways in the gears to ensure correct engine timing.

Extractor used for pulling timing and oil pump drive gear off crank

Bevel gear removed showing drive gear key and shim washers

Removal of the Cylinder Barrel and Piston

Engine on the bench prior to cylinder barrel removal

Undo the nuts that hold the studs in the top of the crankcase halves, thereby releasing the hold on the cylinder barrel spigot. Rotate the crankshaft so that the piston is at BDC. Carefully tap the barrel bottom flange with a hide / nylon mallet to encourage the flange joint to come apart. Use a 3" x 2" block of wood to tap the barrel base in all directions until it starts to move. Patience is required during this operation in order to avoid broken barrel cooling fins. Many barrels can be seen on e-bay and elsewhere that have suffered the same fate. Don't let yours be one of them.

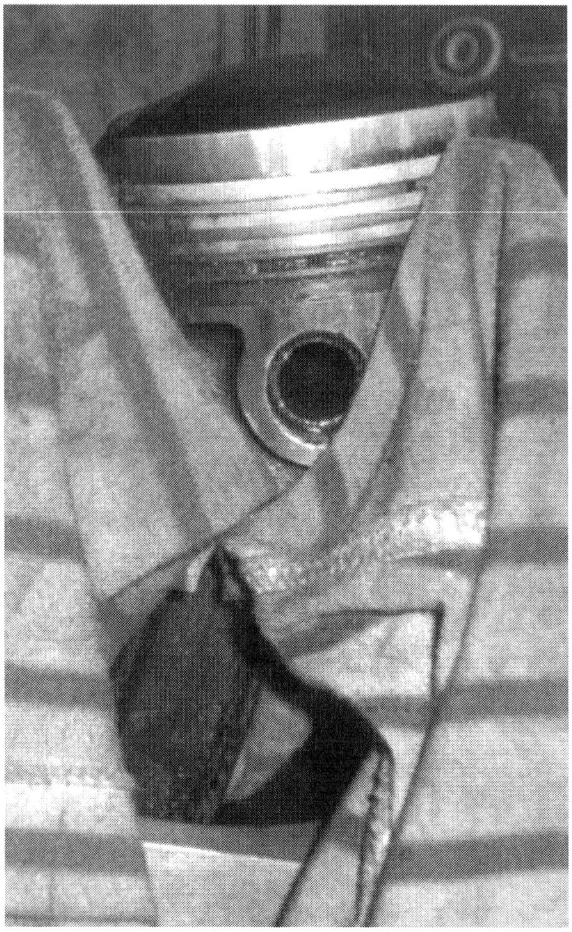

Piston protected by a rag on removing barrel

Lift the barrel so that it clears the piston, while at the same time have a large rag handy to cover the piston and conrod to prevent damage. Put the barrel to one side for safe keeping. Mark the piston with an arrow to designate the front of the engine, then remove one of the circlips holding the gudgeon pin in position. Warm the piston with a hair drier (get the wife's permission first) then push the gudgeon pin through the piston until it just clears the conrod small end whilst holding the piston in the other hand. Put the piston in a safe place; and, more importantly, thoroughly clean wife's hairdryer, and return to the exact position and setting as you found it.

The cylinder barrel on my engine had no broken fins but you can see from the photo where the piston rings had rusted to the cylinder bore. Lots of easing fluid freed the piston rings, but they were in poor shape. The piston cleaned up very nicely, but more of that later.

Cylinder barrel on the bench, note grooves in bore created by rusty piston rings

Splitting the Crankcase Halves

First remove all the crankcase studs and nuts holding the two crankcase halves together, then tap the crankshaft with the hide/nylon faced mallet with gentle persuasion. Also tap on the crankcase joints where one overlaps the other. The seal will eventually break between the two then the drive side half can be separated from the other. The drive side main bearing is a double parallel roller therefore the inner race and the caged rollers will stay on the crank whilst the outer race will stay in the crankcase.

Split crankcase halves

Heat up the timing side crankcase half with a blow torch to just over 100C. sizzling spit temperature roughly, then the single row ball race will come out of the case quite easily with a little gentle persuasion.

Crankcases and flywheel assembly

The main bearings are very special in that they have an internal taper of one thousand of an Inch per Inch corresponding with the same taper on the crankshafts. They are a press fit on the cranks with the slight tapers set so that as soon as the bearings are drawn off the abutment faces they should pull of pretty easily.

Oil Pump, shims and Main Bearings

Now we should set to cleaning the components that have just been stripped. The crank and rod assembly will require a good degreasing and brushing off of all burnt on oil of many years. My crankcase halves internals were full of tar like residues that took some scraping off before proper cleaning could commence. The piston was given a good cleaning and was actually seen to be in remarkable condition with no deep scoring or badly worn ring grooves. More of that later.

The main bearings were given a good degreasing and inspected with a magnifying glass. The races showed signs of wear in microscopic flaking of the hard ground bearing surfaces so replacements were deemed necessary as was expected on such an old engine. The big end bearing was inspected and a little up and down movement was detected along with a discernible amount of sideways rocking movement. This made a replacement big end bearing necessary as well.

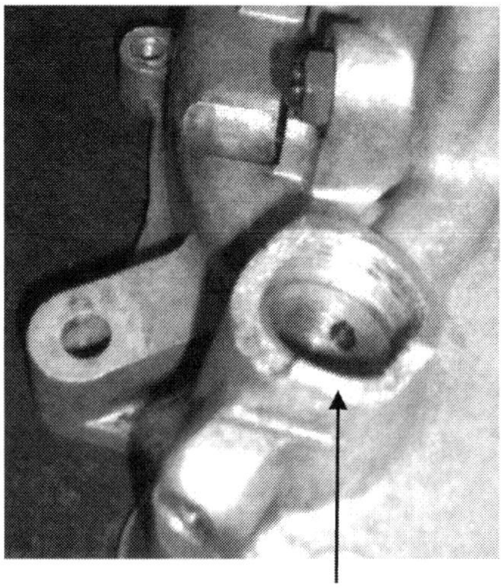

Broken Oil Strainer Boss

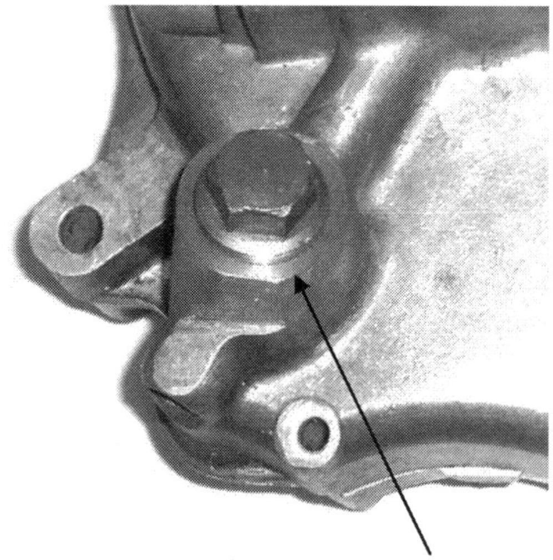

Repaired by TIG welding in new boss

Some damage had been sustained by the crankcases in the dim and distant past. The drive side had suffered some screwdriver damage to the primary chaincase spigot by some amateur "mechanic" and the timing side had a broken boss that held the oil suction strainer. These items were repaired with some very good TIG welding and machining out the damaged screwed boss and making a new one in the lathe and TIG welding in position and then dressed with files just as new. Like all old engines, very little is irreparable, cast iron welding though is a very specialist art and not always successful, hence try very hard not to break fins on cast iron cylinders and heads.

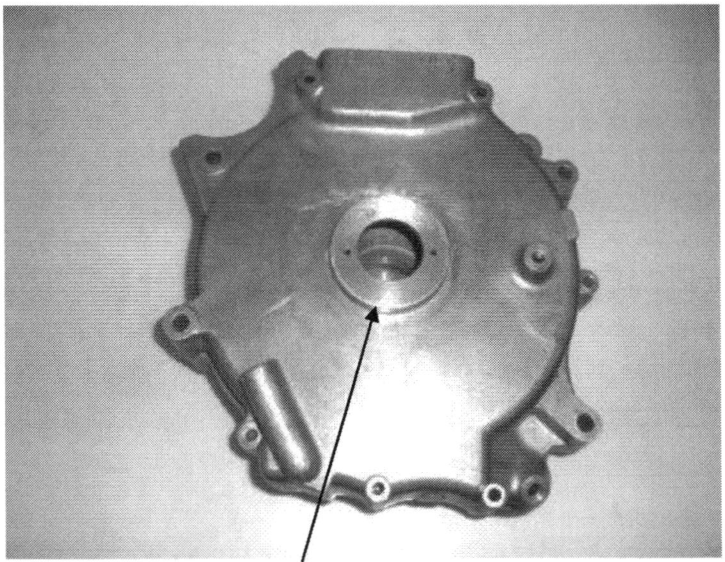

Repaired crankcase bearing boss by TIG welding and finished in the lathe

The flywheels and rod were suitably packed in a wooden box and shipped off to Alpha Bearings who changed the big end and little end bearings, reassembled and trued up, then returned in a large bag of expanded foam.

Flywheels & rod suitably boxed

Flywheels & rod with new big end bearing

Refurbished flywheels (old big end in front)

Cylinder Barrel Inspection.

The barrel was 0.030" oversize and, with the corrosion from the piston rings, would not clean up. It was decided in this case that the cylinder bore would be resleeved. I bought a new piston but, on weighing the new and old pistons, the new one was much heavier than the old. Acting on the advice from my engine machinist, we decided to use the old piston as micrometer measurements and close inspection with a powerful magnifying glass revealed that the old piston was hardly worn at all.

Time in storage had taken its toll on the piston rings. I deduced that the engine must have had a rebore probably not long before being taken off the road for the last time, hence the poor condition of the rest of the engine. The new piston was returned and the new cylinder sleeve pressed into the bored out cylinder, then the liner was finish bored to suit the 0.030" oversized piston now fitted with a new set of piston rings.

Resleeved cylinder barrel

Checking piston ring gap

Reassembly of Crank and Main Bearings into Crankcases.

The crankshaft has to be shimmed either side of the main bearings to; first, centralise the conrod in the cases; and, secondly, to remove all discernible end float in the crankshaft. This is essential for correct engagement of the camshaft drive bevel gears for constant minimum backlash between the bevel gear teeth. This is a very skilled operation and should only be attempted by those with the essential knowledge and tools and equipment to do the job properly. One thousand of an Inch end float of the crankshaft is the maximum permissible.

If you have a fully equipped workshop with a good array of internal and external micrometers, one can calculate the shims required from measurements taken of the crankshaft and the distance between the two main bearings when the crankcases and bearings are assembled.

Flywheel assembly ready to be fitted into crankcase halves suitably shimmed to limit endfloat

Main bearings assembled and crankcase halves sealed to make oil tight

The alternative method is to fully assemble the crankshaft with new main bearings into the crankcase halves with no gasket, first ascertaining that the shims initially used will give some backlash in the crank when measured with a dial test indicator on the crankshaft end. This measurement will determine the adjustment to be made of the shim pack. The shims always go on the outside of the bearings, not between the bearings and the flywheels. Remember the slight taper on the crankshaft and bearings. On the timing side, the outer shim is an oil retention washer that prevents pressurised oil in the lower bevel casing entering the crankcase.

When all shims have been set, reassemble the crankcase halves and crankshaft with no gasket but a thin film of sealant between the flange faces which must be spotlessly clean and free from imperfections to ensure an oil tight engine. Note the essence is a THIN FILM of sealant. Silicon sealant if allowed to get into oilways can restrict oil flow. So if you have a large gap to fill, do not use sealant to bridge the gap.

Dial test indicator

Final checking of crankshaft endfloat to ensure maximum of 0.001"

Remember that when assembling the main bearings into the crankcases, the cases must be heated to just over 100C. Sizzling spit temperature. The bearings will slide into the crankcases at this temperature. Do not overheat the crankcases or distortion will take place.

Before fully tightening up the crankcase studs, make sure that the cylinder barrel face two halves on the crankcases are perfectly level with each other.

Oil Pump

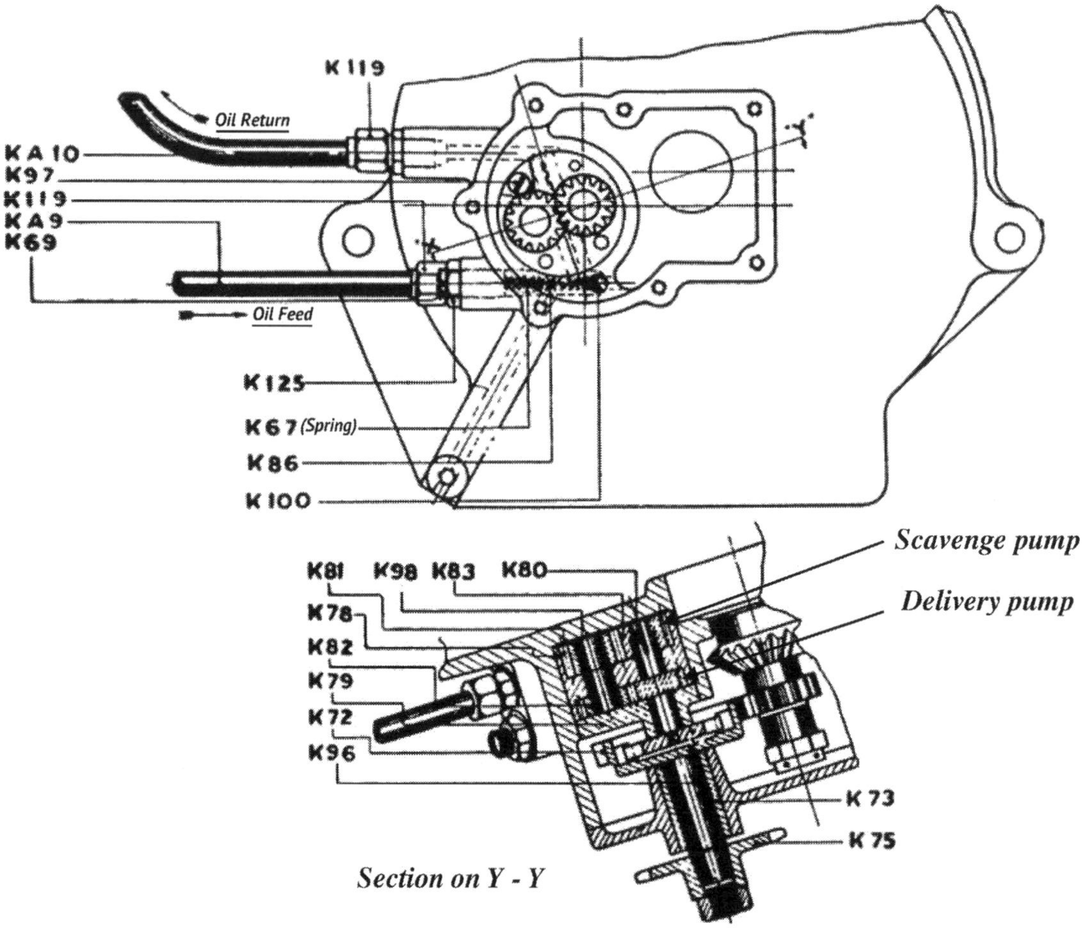

View of Oil Pump with Timing Case Removed & Sectional View of Oil Pump

Oil pump removed from the crankcase

The oil pump is the heart of the engine. Without it operating at its design pressure and flow rate, the engine will not give of its best in terms of life span and performance. The pump is, in fact, two pumps in one, driven off a common shaft. One is the delivery pump supplied with oil from the oil tank; the other is the scavenge pump which draws oil from the crankcase and returns it to the oil tank. The pumps are virtually identical, being of the geared variety, the scavenge side being twice the width of the delivery side, thereby ensuring that the crankcase never floods whilst the engine is running.

The Velocette oil pump is a precision bit of kit which deserves full attention on a rebuild and therefore will require removal from the crankcase and a full inspection and replacement of any worn parts. It is partly protected by a suction strainer on the scavenge side, and a strainer in the oil tank for the delivery side.

The pump is driven off the crankshaft on the timing side by a spur gear in front of the lower bevel camshaft drive. This gear drives from the rear of the magneto drive shaft which contains an internal Oldham type coupling to compensate for inaccuracies in the drive line. This coupling consists of a small circular plate with two tongues on the outer edge that locate in slots inside the spur driven gear. A slot is machined in the plate at 90 degrees to the outer tongues which drives the oil pump shaft through corresponding flats machined onto the shaft. If the pump shows signs of excessive wear and/ or corrosion, bin it and buy a new one. They are not cheap, but neither are the parts for a complete engine rebuild. Gear sets are sold in pairs but the pumps are normally sold as a complete assembly.

Recesses for inverted "Oldham coupling"

Timing gear in timing case with recesses to drive oil pump via coupling below

Oil pump drive inverted "Oldham coupling" that sits in slot in gear on previous page

To remove the oil pump from the crankcase without stripping the whole engine is a protracted business, requiring much patience and precision. The crankcase half containing the oil pump requires localized heating to about 100C and, with the fixing screws removed, two of the holes in the oil pump body require tapping out to ¼ BSF for a depth of 3/8 inch to take two screws used with an extractor to remove the pump. Attempting this operation without heating the crankcase and thus expanding the surrounding aluminium hence relaxing the transition fit will destroy this fit that enables the pump to deliver oil at pressure. The factory assembly method is as described in the Velocette instruction book and that is with the engine totally stripped and the pump removed at the same time as the main bearings are removed. Heat the crankcase in the oven to just over 100C (sizzling spit temperature) and with the oil pump retention screws removed, take the hot crankcase to the bench and tap down sharply the timing chest face onto a solid piece of wood whereby the oil pump will just fall out. Never heat the crankcases above 160C or distortion will occur.

Take note as to how the pump is assembled before removal then you will know later how it goes back together. The pump gear will come apart and after cleaning all components in petrol, carefully inspect the gear teeth and the closeness of the meshing of the gear teeth. Any imperfections in the surface finish of the gear teeth rubbing surfaces such as undue scoring and grooves will be indicative of foreign bodies passing through the pump and causing unrepairable damage causing loss of oil pressure. The end faces of the gear teeth and the brass shim at the base of the pump should also show little signs of wear. If any doubt about the serviceability of the pump, for peace of mind, fit a new one.

To reassemble the pump into the crankcase, fit two $^{3}/_{16}$ Inch dia x 3 Inch long studs, with slots cut in the stud top to remove with a screwdriver. Heat up the crankcase as before for removal then insert the fully assembled oil pump with brass shim over the two studs smartly and tap down with a brass drift to ensure the pump body bottoms out in the crankcase. Insert two of the countersunk $^{3}/_{16}$ Inch screws, remove the two studs, and then insert the remaining two screws and tighten down firmly to secure the pump in position. Ensure that the oil delivery slot in the pump outer case is at the top. Lubricate all moving parts in engine oil and insert plastic or rubber blanking plugs into the suction and delivery unions to keep out any debris.

Oil pump replacement with two studs to align screws

Bevel gear and timing pinion assembled.

Note oil pressure spring and ball in crankshaft

Cylinder Head

The cylinder head should already have been cleaned in petrol to remove all oil, grease and dirt. The cam drive gear and camshaft stripped and laid to one side for further cleaning and inspection. The cam is drawn off the drive shaft with an extractor. The drive shaft is removed from the top bevel housing by first heating up the aluminium which expands the metal allowing the shaft to be withdrawn. Do not attempt it cold. The shaft diameter at the cam is larger than the bore, to give a press fit in the cam.

The rocker shafts, rockers and associated shim washers are then removed, followed by the valve springs, cotters and finally the valves. The valve spring seating washers are then removed revealing what is left of the valve stem seals.

Camshaft and Rockers prior to re-profiling

Valve/ Springs & Cotters Rocker/ Spindle/

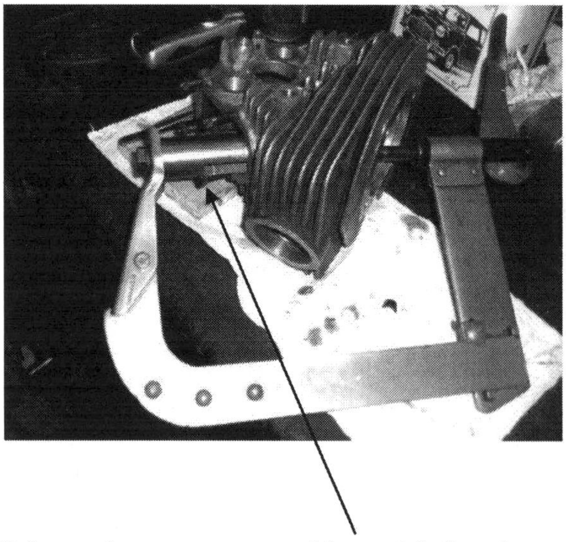

Valve spring compressor with special slotted adapter to access spring cotters

Using tweezers to remove spring cotters

Fully stripped and cleaned cylinder head *New valves being inserted into head*

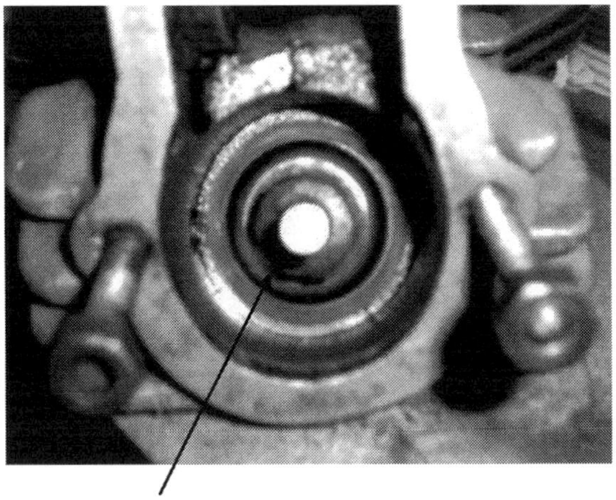

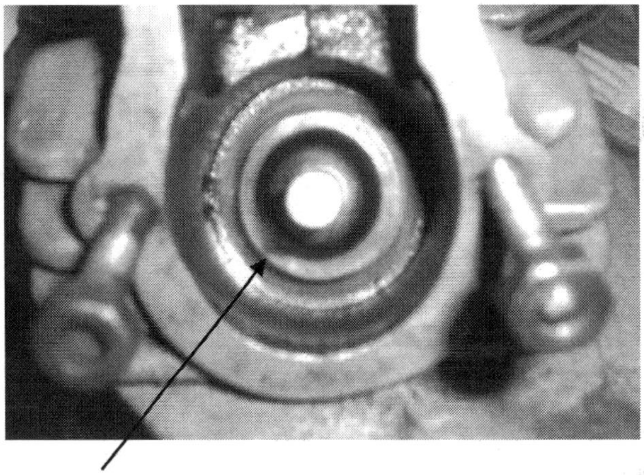

Leather seal placed over valve stem *Top hat placed on top of valve stem seal*

The valve spring seals are, in fact, made from leather as original. I had thought of using modern stem seals, but could not find anything suitable. I then considered using O rings but was concerned that this might starve the guides of oil to the detriment of the valves. So I settled on making my own leather seals from an old belt.

The original valve seat inserts valve seating faces were badly recessed, so these were removed and replaced with modern sintered iron alloy (non leaded petrol resistant) commercially available insert blanks. These were machined to suit the remachined insert recesses in the inlet and exhaust ports. The heavy interference fit necessary to prevent the seats from dropping out at running temperatures was adopted at the machine shop. Assembly of valve seat inserts into the head was done by heating the head and freezing the inserts in liquid Nitrogen. Insertion is a skilled job, as you only get one chance at getting them in properly.

The seats were then cut concentric with the new valve guides previously fitted.

New inlet and exhaust valves were then lapped in using fine carburundum paste, washed out in petrol and thoroughly cleaned to ensure the removal of all abrasive material.

The cams and rockers showed signs of excessive wear, and, in the interest of future reliability and performance, it was considered worthwhile having them reprofiled by experts. The worn cams and rockers were entrusted to Newman Cams for reprofiling.

Cams removed from camshaft ready for re-profiling (Note woodruff key)

Re-profiled cams assembled on drive shaft

Finished top bevel drive assembly

The full range of cams used by Velocette throughout its production of nearly all of their road and race machines are shown in the table in the Technical Data section. This table gives full details of cam valve timing and also ignition timing for each model inclusive of all OHC models and pushrod machines up to the MSS, and the LE side valve. This table appears to be quite accurate, but is only my assumption based on differing figures I have been quoted for the KSS Mk2. I have found the ignition timing figures for that model in this table to be the most suitable for my bike from experimentation between 30^0 to 40^0 BTDC and then backing off to 38^0.

Having established that the cams in my KSS were the K17/10 having had them reprofiled, I found by using a large engine timing disc that the cams opened and closed the valves at the right angular positions with respect to the other cam. To get the valve timing correct to the book, I had to set the timing disc in the correct angular position to indicate TDC with a pointer attached to the engine.

Setting the timing disc and pointer was achieved by the time honoured 'positive stop' method with the cylinder head removed and a ground stock steel level bar to stop the piston as it approached TDC.

Determining TDC using a straight edge as piston dead stop

The timing disc was rotated to give equal angle either side of TDC. The cylinder head was then replaced and cam timing commenced. A first attempt was made by following what was thought to be the timing marks on the bevel gears. On closer inspection there was found to be more than one timing mark on the camshaft bevel gear. I had to ascertain which was the best mark to follow by centralizing the fixing studs in the slots in the bevel wheel, and finding the best timing tooth mesh to achieve the nearest timing combination. On finding this I then used the adjustment provided in the slotted holes to adjust the camshaft bevel to give the nearest valve opening and closing angles as possible. Taking into account manufacturing and build tolerances, I arrived at a camshaft timing I was comfortable with. I then etched my own timing mark on the bevel gear with an acid etch pen.

Subsequent running of the engine and the bike on the road have confirmed that I was not a million miles away.

I have shown a photo below of a KSS road cam alongside a KTT race cam which clearly shows the longer valve opening duration of the race cam required for sustained high speed running.

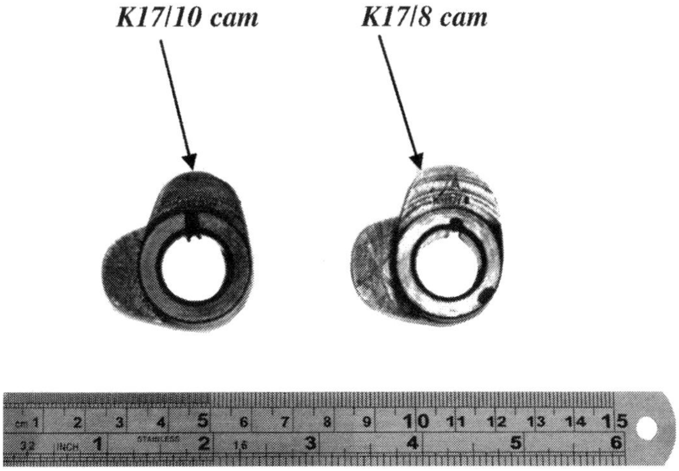

Comparing cams K17/10 later KSS Mk2 & K17/8 KTT MkVI

(Note the long opening time duration of the race cams – not recommended for road use)

Carburettor

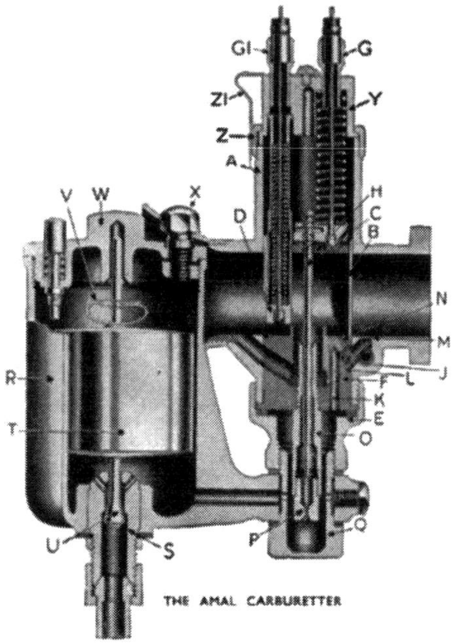

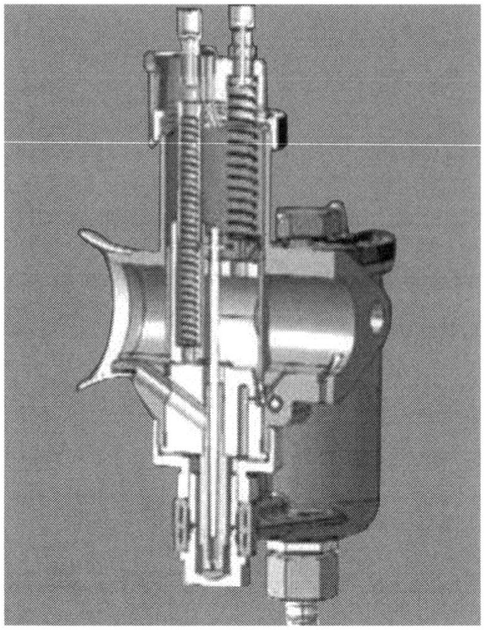

Cross Sections through the Amal Carburettor

The Amal carburettor that came with the bike was indeed confirmed as the correct one for the year of the bike after reading up on extracts I had from old Amal catalogues. Later confirmation was given after consulting Amal. 1948/1949 was a change point in this particular Amal specification to increase the bore of the carburettor from 1" to $1^{1}/_{16}$". The carburettor designation is Amal 276CV/1J as against 276BU/1J for the earlier model.

The carburettor was in a sorry state so was subjected to a complete strip down and clean up. It was well oxidised externally, but appeared serviceable internally. I gave it a good wire brushing externally and washed it thoroughly in petrol, giving all the internal surfaces and petrol passages a good cleanout. On close inspection it became apparent that the air slide was badly worn and rattled in the body indicating the necessity for a replacement slide at least. I looked into the cost of a replacement carburettor from Amal and other suppliers, and then decided to renovate the original one.

The jet block was removed from the carburettor body by gentle blows with a lump hammer and an aluminium drift. Then the whole carburettor body and jet block were subjected to a blast and soaking in carburettor cleaner. This ensured that all the minute drillings for the pilot supply were well and truly clear. With the aid of an inspection torch, I could see light through the small drillings confirming that no obstructions remained.

I purchased all new jets, needle, air slide, springs and seals, etc from Amal and assembled the carburettor with refurbished and replated chromed bits like the air trumpet and other plated items, which revitalised the complete assembly. The new air slide was supplied over size by Amal who gave me instructions on the best way to make a good sliding fit in the carburettor. This I did with varying grades of wet and dry paper and the aid of engineers blue to highlight the areas to be removed. With a lot of patience I arrived at a lovely sliding fit which is essential for good throttle operation along with controlling any air leaks past the slider, particularly on tick over and small throttle openings.

 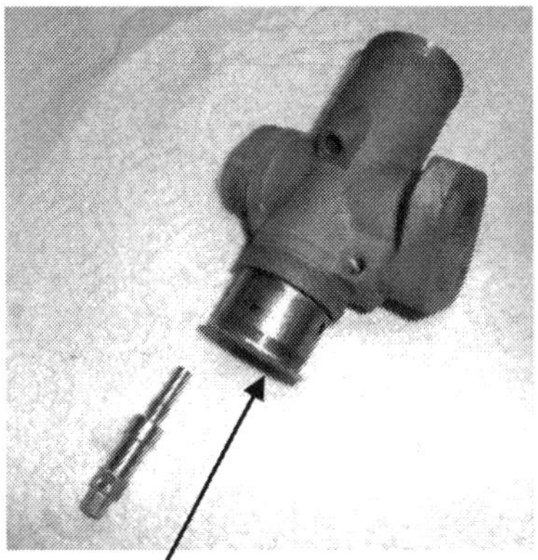 

Carburettor stripped for cleaning   Jet Block pressed out to reveal...... idler jet orifices

Assembled onto the engine, it really looked the part and saved a considerable sum over the price of a new one (see later paragraphs). It must be emphasised though that Amal do a remarkable service in manufacturing new carburettors for all classic British bikes and provide an excellent service in spare parts. Their service manuals and parts books are very well put together and serve as excellent reference sources for the settings for all our machines.

The use of any other make of carburettor whether it be period or not, spoils the look of an old classic. A new pattern of carburettor may breathe new life into an old engine and enhance the slow running capabilities and overall fuel consumption, but does not go in keeping with the appearance of the original.

Extracts from Amal Tables

MAKE	MODEL	ENGINE	YEAR	SPEC NUMBER
Velocette	KSS	348cc O.H.C.	1946 to 1947	276BU/1J
Velocette	KSS	348cc O.H.C.	1948 to 1949	276CV/1J

Carburettor Specification: 276CV/1J

Make:Velocette Model:KSS   Year:1948 to 1949   Engine Size:348cc O.H.C.

Number of carburettors:1   carburettor Type:Type 276

Internal Bore: $1^{1}/_{16}$”   Needle Position:3

After some time, it became apparent that the old refurbished carburettor was indeed proving difficult to achieve a constant low running speed. Maybe internal corrosion that I could not see had taken its toll. Whatever was causing it, I took the decision to purchase a brand new carburettor from AMAL. This was an alternative to having the old carburettor completely renovated including sleeving. As this was not guaranteed to completely cure the problem, I considered it safer to buy a new one. This did, in time, prove to be the right decision.

The strip down information I have outlined for the old carburettor is still valid if you consider your old carburettor to be serviceable.

The refurbished carburettor
(Looked the part but was not sympathetic to fine tuning)

New Amal Carburettor as fitted

Drive Train

Gearbox

The gearbox came complete with the bike with the end cover as a loose item. I had checked that everything appeared to be there as we put it in the box with the rest of the bike in other boxes or loose as the case may be. The only piece missing from the whole bike was the gear lever and associated clamp. I made the lever later from stainless steel tube and purchased the clamp from a chap who had advertised in the Fishtail magazine.

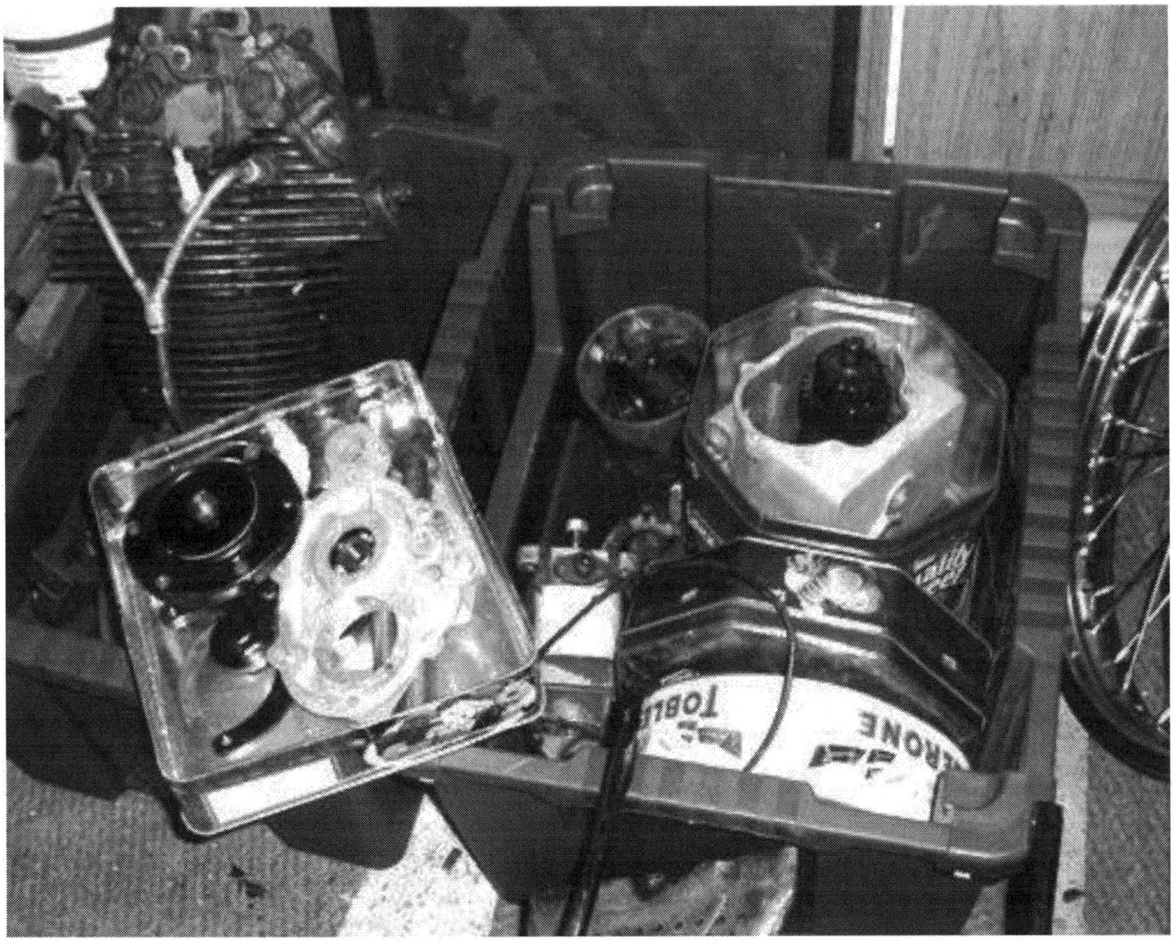

Engine, gearbox and other items as brought home in boxes

Having removed the gears it became apparent that the gear teeth had suffered quite a lot from corrosion. The old Velocette malady of water entering the box from the clutch cable entry into the top of the gear box had taken its toll. If the oil level in your Velocette gearbox is increasing, suspect water entry and drain it off and replace with fresh oil immediately!

A rubber sleeve can now be purchased that fits over the clutch cable at the gearbox end that prevents water from entering the gearbox.

Top main shaft

Lower lay shaft

Main shaft gear set not totally a disaster but the lay shaft set showed signs of being submerged in water rather than oil.

I stripped the gears down to their individual components, cleaned them up as best as I could to remove the rust with a fine wire brush, then submerged them in phosphoric acid (drain cleaner) and watched the rust dissolve.

I built the gears up again and assembled the box with new bearings all round. I was never happy with this as the best solution and searched high and low for replacement gears.

A new set of gears purchased at the Stafford Show

Eventually I came across a good set at the Stafford Motorcycle Show and replaced them along with other components from the same source such as indent pawl, cam plate and selector forks which were in better condition than my originals.

The Velocette gearbox is quite simple in its design and is relatively easy to strip and rebuild. It can be stripped in the frame, but better on the bench where it is easy to handle.

The procedure is commenced by removing the starter shaft complete with attached lever. If no fault is suspected such as a broken or weak return spring then it is best left alone. Do not misplace the thrust washer with the three thrust pins behind.

Then unscrew the blanking cap that holds the main shaft outer bearing in situ. If the box is still in the frame with primary drive in place, then lock up the gearbox by engaging the rear brake with the box in top gear. If the box is loose on the bench then clamp the main shaft by attaching the output sprocket on the cruciform end as is its normal drive, and tighten up the nut and washer to hold firm. Using a length of final drive chain or other torque resistance device such as a vice, undo the nut on the other end of the shaft. Remove the nut and washer and put in a safe place.

The next stage is to remove the two nuts and washers off the ends of the selector shafts. With a hide mallet tap the main shaft partially through the end casing bearing. Having done this, the end cover of the gear box can be removed. With the aid of steel bars across the gear box face, and with the nuts and washers screwed back on the selector shafts, prise out the selector shafts from the rear wall of the box with levers. The selector shafts are pressed in the casing with a light press fit so should come out relatively easily.

The gear shafts can now be removed along with the selector forks, after removing the centre main shaft. The KSS main shaft gear set has to have the gear teeth arranged in a specific way to allow the layshaft gears to become disengaged. This will become apparent by trial and error but note this position to aid reassembly. MSS gear sets do not have this problem due to slightly different gear ratios. i.e. 16 teeth versus 17 teeth on the sleeve gear.

Inspect the gear change positive stop mechanism for correct functionality but dismantle if found necessary to replace any worn components. Worn pawls and weak springs are the only items that require to be replaced unless like my own box you find that water in the box has caused corrosion damage. The only recourse at this stage is to replace with new or salvaged useable items as found necessary.

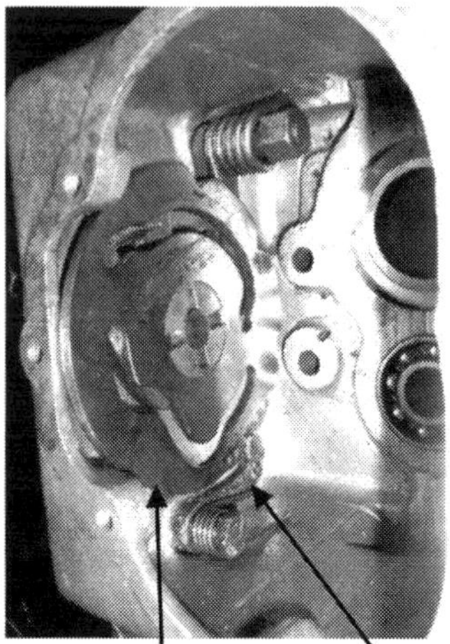

Corroded camplate & pawl

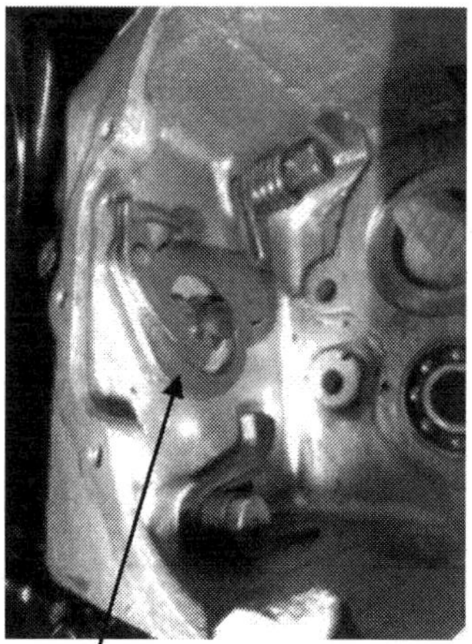

Replacement mechanisms

Replacement camplate & pawl

Original gear change mechanism badly corroded, replaced with perfectly serviceable components

After ensuring that the gear change mechanism is in perfect working order, the first item to install is the mainshaft bearing from the outer side of the casing with new dirt shield and correctly screwed down retaining screwed ring. Ensure that this is staked in over the notches to make sure it does not unscrew in service. Replace the lay shaft bearing in the casing. Replace with new bearings if the old ones show any signs of wear or just plain rumble.

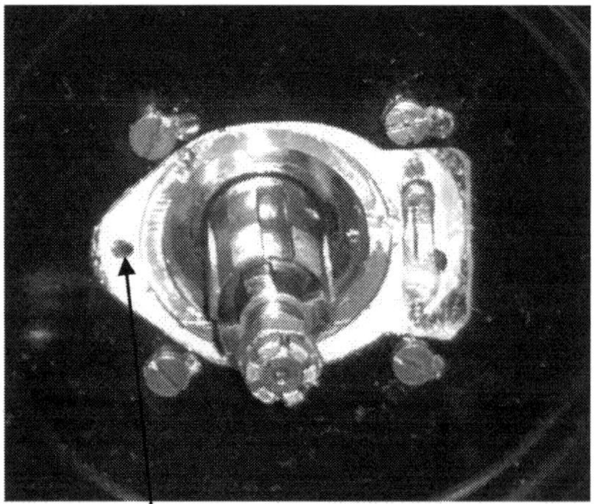

Clutch operating pin

Clutch operating release mechanism fulcrum

At this stage, check that the clutch release mechanism inside the gear box operates correctly whilst it is still accessible. Replace any suspect components but these are low stressed items that very rarely are cause for concern.

First gear to go in the box is the sleeve gear along with the oil flinger which must be the right way round. Check the manual.

Then, holding the gear sets in both hands along with the gear selector forks, insert the gear sets into the gearbox casing followed by the mainshaft through the sleeve gear from the primary drive side of the box into the main gear set. This may sound fiddly; but, after a few attempts, the gears will go in. After doing this a couple of times it gets easier to do. Locate and tap in the selector shafts that will make the selector forks more secure. Ensure that all the thrust washers that came out of the box go back in.

Before assembling the end casing onto the box, make sure that the gasket faces are clean and flat. Attach a new gasket with jointing compound spread evenly on both faces, and assemble the gearbox end casing onto the box along with the selector shaft forks. Torque the bolts up evenly and assemble the nuts and washers to the selector rods, tightening them up evenly.

Oil Filler orifice and level point

Assemble the kick start assembly to the end of the gearbox with suitable sealant but note there is no gasket at this joint. Make sure that the thrust washer at the end of the layshaft is in position. Position the three bolts noting that the long bolt is at the lowest position as it goes through the gear box end casing into the gearbox shell. Seal this long bolt washer face with sealant as oil can leak out along this bolt. After allowing jointing compound/sealant to set, fill the gearbox with oil up to the oil filled level screw hole. When oil appears at the hole insert the oil level screw. Screw the oil filler plug into the end casing with a new fibre washer.

Test the gear change mechanism by rotating the input and coaxial inner output shafts whilst operating the gear change lever at the back of the box to check for correct operation of all four gears and the neutral position between first and second gear. Check for smoothness of operation and smooth running of the gears.

Primary Drive including the Clutch

To disassemble the primary drive including the clutch on a Velocette single is not as straight forward a task as on most other British motorcycles from the same period. This is because of the unique way that the final drive sprocket is positioned outboard of the primary drive sprocket and clutch assembly. This is mainly because of the very narrow proportions of the crankcase that make an inboard primary drive possible. This has obvious disadvantages as far as dismantling a clutch and changing primary chains is concerned, but on the other hand makes for easier changing of final drive sprockets and chains which is a boon to short circuit racers who need a wide selection of final drive ratios - swings and roundabouts. The whole Velocette design concept is steadfast adherence to solid engineering principles viz (1) compactness for reduced frontal area to minimize air resistance, and (2) enabling low mounting of the engine / gearbox to reduce the C of G for good handling, whilst retaining adequate cornering clearance. Keep these concepts in mind as they will see you through the workshop hours dealing with the downside of reduced access to the primary drive and clutch.

Velocette single cylinder engine clutches are a master piece of engineering if assembled and maintained correctly. If not they can be cause for untold misery. Unlike most other motor cycle clutches, Velocette clutches do not operate with a central coaxial pushrod running through the centre of the gearbox main shaft. The mode of operation is more like a car clutch where the operating motion for the clutch plate separation mechanism is made external to the shaft at one side. This operating mechanism takes the form of a lever between the rear of the clutch and the gear box casing pivoted at one side of the shaft and moved away from the gearbox casing by a pin penetrating the gearbox casing which in turn is operated from inside the gearbox via a bell crank and pull rod connected to the clutch cable at the top of the gearbox.

This pin operated lever has the gearbox shaft running through it with a thrust bearing transmitting the axial forces through the clutch to separate the clutch plates. This thrust bearing is maintained parallel to the clutch mechanism by virtue of a spherical face at the rear of the clutch release bearing and a corresponding spherical seat in the operating pivot lever. See pictorial representation of clutch operating mechanism on page 84.

Clutch disassembly

Access to the primary drive requires quite a bit of initial work in order to access the main components. First, remove the rear brake lever then remove the dynamo and associated drive after taking off the dynamo belt drive cover and removing the dynamo pulley. Bend the tongues of the crankshaft nut split pin straight and remove with long nose pliers. With the aid of the special castellated ring spanner, undo the crankshaft nut and remove along with the dynamo drive pulley. The "cush" drive spring can also be removed at this point. Next remove the left hand footpeg complete with inner nut and washer from the footpeg square mounting spindle. This inner nut will impede the removal of the outer chaincase.

Remove the rear chain guard front piece that obscures the rear of the primary chaincase. This usually involves partially dismantling the rear section as well. At this point, the primary chaincase can be dismantled. After doing all this, one can understand why the Velocette clutch has to be assembled correctly and adjusted in the sequence as devised by the factory. When new components have been used along with new clutch friction plate linings, adjustments have to be made when everything has bedded in. These adjustments can be made without the dismantling procedure shown here and are fully explained in the workshop manual. More of this later.

Before the primary chain case can be removed, the gearbox final drive sprocket has to be removed along with the rear chain. Release the rear wheel spindles and undo the rear wheel adjusters in order to move the rear wheel forwards thereby slackening the rear chain. Remove the split pin from the castellated nut holding on the gear box sprocket. Undo and remove the nut and washer which will probably require the rear brake pedal being reinstalled temporally. Remove the sprocket which is located on the inner gear box shaft which has a cruciform drive arrangement. If the nut has not been tightened up properly in the past, there may be wear in the cruciform slots in the sprocket. The shaft is much more wear resistant. If the sprockets and chains have seen much service, they will have to be replaced. Hooked sprocket teeth and sideways bendy chains usually confirm this.

Now the primary chaincase can be removed. There is no oil drain point on the KSS chaincase so have a suitable drip tray handy to catch any oil in there. Oil tight primary chaincases on Velocettes are a rarity, but if the engine has been subject to excessive blow-by into the crankcase then some oil will find its way into the chaincase via the left hand main bearing.

Before dismantling, clean up as much external dirt and oil from around the chaincase to prevent ingress of harmful extraneous matter into the working parts. There will be enough muck in there without adding to it.

Primary chaincase tinware held together with 14 No.– 2 BA screws. Narrow gasket faces make sealing very difficult, not counting for the large holes in the front case.

Firstly, clean up all the screwdriver slots in the 14 No. 2BA screws that hold the chaincase halves together. Armed with a good straight screwdriver of a width to suit the screws, unscrew each of the screws a little at a time for the first couple of turns, then carefully remove and place in a tin along with other storage containers which will be required to ensure that nothing is lost or misplaced. At this point, I find it useful to have a stock of various sizes of sealable polythene bags that can be marked with a marker pen what the bits are and where they came from. Seems a simple system but believe me, it works.

If the chaincase halves are well stuck together, be careful not to use excessive force to separate them as distortion of the flange faces will result and it will never seal effectively ever again. Use two screwdrivers, and tease the outer chaincase to separate from the inner from the top and work gradually round until they become to part company with one another. As you approach the bottom of the chaincase it will be found that clearance with the frame bottom tube is very small so be careful not to damage the paintwork.

Any excess oil and sludge in the chaincase will have deposited itself all over the garage floor (or the kitchen floor much to your partners annoyance) if you forgot to place a drip tray underneath. The outer chaincase has to be juggled over the engine main drive shaft and the clutch and final drive shaft. Clean everything up with lots of rags and cloths and place the chaincase in a safe storage place. The contents of the primary chaincase can sometimes give a clue to the extent of abuse and neglect that has been used in the past by the state of the sludge, bits of broken chainrollers, dead mice, etc, etc.

The clutch and engine sprocket, complete with primary chain have to be removed together. There is no other way. First lock up the gearshafts as previously, then remove the locking tab from the sleeve nut at the centre of the clutch. Undo the sleeve nut with a peg spanner as provided by Velocette as part of the "tool kit". A much better tool is the factory designed tubular special tool with the hardened pins in the end, used with a tee bar. It is a much more efficient tool for undoing and tightening up this much maligned item. Damage will result if the clutch assembly is not tightened to a sufficient torque to prevent it rattling around on the shaft,

Locking tab on the sleeve nut to be removed first by unscrewing small countersunk screw with a small screw driver.

Another useful tool is the large castellated double ended spanner that serves both the clutch spring carrier and the crankshaft main slotted nut. It has a large slot in the handle that goes over the crankshaft to prevent the clutch spring carrier rotating while making adjustments. This is an aftermarket tool which I was thinking of making something on similar lines only to find that someone had beaten me to it. A good investment if you want to keep the adjustment slots and crankshaft nuts in good condition and prevent bleeding knuckles. Hammers and chisels are not recommended tools for Velocettes. Hammer and chisel rash on any part of a Velocette is a bad sign, as are chewed up nuts and bolts and deformed screwdriver slots in screw heads.

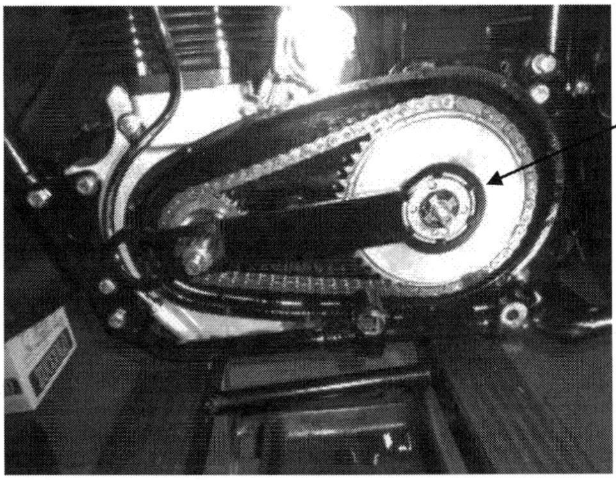

Special spanner used for holding spring carrier in position

Tubular pin spanner with Tee bar for undoing and tightening up the clutch sleeve nut.

In order to dismantle the clutch assembly it becomes essential to loosen up the spring carrier in situ by the use of a 'C' spanner and an aluminium plate sprag between the two sprockets to prevent them from rotating. With everything loosened up, the sleeve nut can be finally unscrewed and the clutch assembly removed from the gear box shaft in conjunction with the engine sprocket and primary chain. That is as described in the Velocette Instruction Book, price 2/6d. Do not believe it. I have found it necessary to dismantle the clutch piece by piece in situ and when you get down to the middle plate which carries the large primary chain sprocket on the carrier bearing, then the primary chain and engine sprocket can be removed as one. Here again, another obstruction comes to play in the bracket welded to the frame that holds the left hand foot peg. In order to get round this, the gearbox has to be moved forward to give enough slack in the primary chain to get the chain over the bracket. Try doing all that with one pair of hands. It can be done but extremely difficult. Another pair of hands makes it much easier.

Reassembly is slightly different in that the fully assembled clutch can be held together with tie wraps (not available in 1947). The Instruction Book and other manuals recommend a spare sleeve gear to assemble the clutch on the bench and hold together with string. I did not have a spare sleeve gear at the time and made do with sight of eye and some engineering skill, and the tie wraps to hold it all together whilst everything was tightened up.

Aluminium plate sprag

'C' spanner

Use of 'C' spanner and aluminium plate sprag is required to prevent turning of the primary chain system.

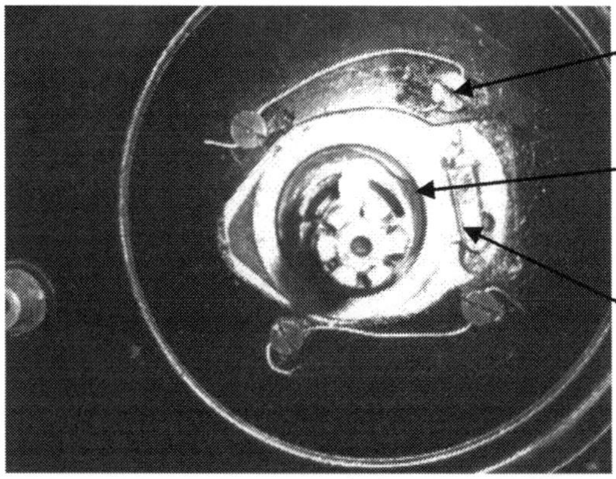

Primary chaincase held to gearbox with four shoulder screws wired to prevent unscrewing.

The spherical contoured thrust washer is contained within the counterbore of the banjo housing clutch operating gate.

The gate pivot is held up to the pivot plate by the wire retainer held to the gearbox casing by two small screws. This has to be disassembled in order to proceed.

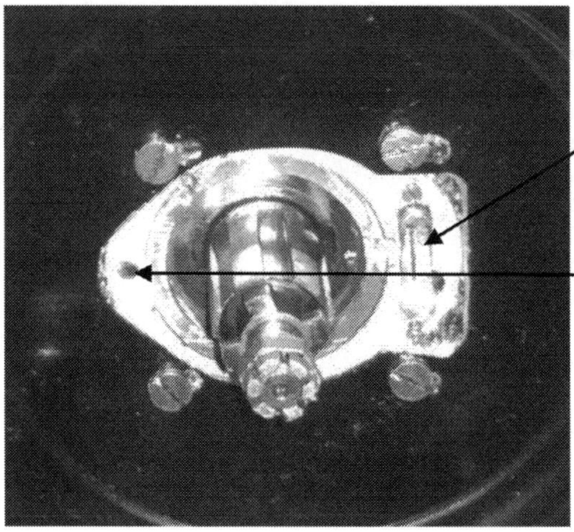

The plate to the right of the shaft is the pivot point for the clutch operating gate.

The operating pin is to the left of the shaft. On disassembly, be careful not to loose this pin.

Clutch Inspection and Relining

Any damaged or distorted clutch plates should be replaced; and, if any of the friction plates inserts are worn down to the steel surface, the inserts will need replacement. There are 22 inserts in each friction plate. There are three of these, including the centre plate that holds the large sprocket and the carrier bearing. Either side is a smooth steel disc or plate including the back plate, the front cover and two thin inner plates.

If the centre plate sprocket is damaged in any way or teeth worn as shown by a hooked appearance, then it must be replaced as should the engine sprocket and primary chain. Also, if the clutch carrier bearing is worn and noisy, it should be replaced. They can become loose in the housing and if this is the case a new bearing may correct the press fit required but if not, a new sprocket and clutch centre will be required. It must be said though that these components are quite hard; and, if the primary chaincase is kept well oiled, wear in these components is rare.

Modern Velocette clutch replacement parts can be specified with sintered linings but how these perform in older machines I would not know. They do make replacement more difficult in that new parts have to be purchased rather than just friction inserts.

This design of clutch derived from the earliest days of Velocette from the earliest two strokes and the 250cc MOV which relied on only three plates! Increased torque capacity has evolved over the years by adding more plates and springs. In comparison, the Venom Thruxton clutch is a bit of an animal by all accounts.

The KSS clutch, set up correctly, is as good as any modern variant I have come across. Apparently once set up they can go on for very long periods without attention. Not many old British bike clutches could carry that accolade, but there is a sting in the tail. If they are not set up correctly in the first place or are incorrectly adjusted in service, then all sorts of maladies can ensue. A dragging, slipping clutch all at the same time is a common fault and has given Velocette clutches a bad name in the hands of people who do not understand how they work.

Clutch components laid out on the bench ready for inspection

To reline the clutch friction plates it is necessary to punch out the old inserts. Care is required to avoid distorting or bending the plates.

New inserts are readily available, 22 in each plate

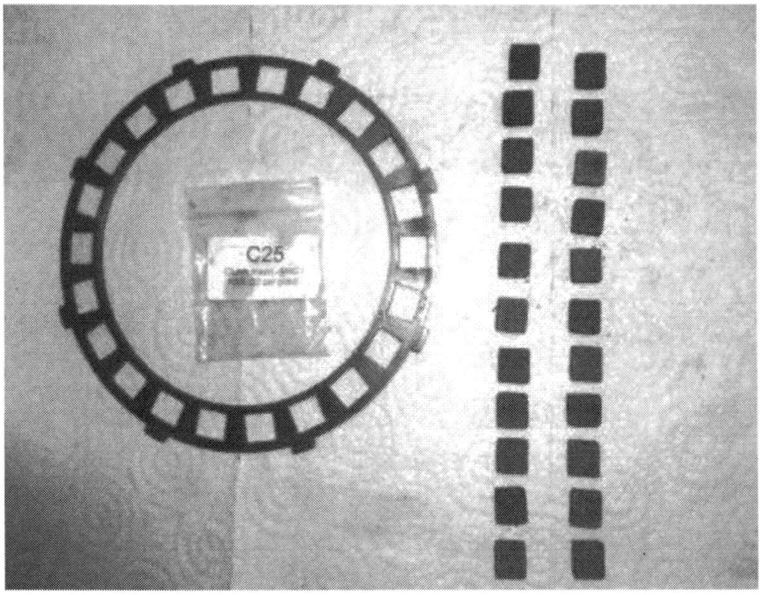

New clutch "Ferodo" inserts ready for insertion into clutch plate

Take each empty clutch friction plate in turn and lay it on the outer flat plate. Take a clutch "Ferodo" insert and carefully press into a square hole with your fingers. Then holding the plate up in both hands, manipulate the insert so that it is sticking out equally on both sides of the plate. Lay the plate back down onto the outer plate and repeat for the other 21 inserts. Do this for the other friction plate and also the centre plate. When this has been done, pour a small quantity of auto gear box oil into a tin just large enough to hold the clutch. Submerge the plates for 24 hours. This will make them expand and fit tightly in the plate recesses. It also enhances the performance of the clutch overall as in time the plates will become contaminated in primary chaincase oil and a clutch run wet to begin with is reputed to give better overall performance than one run dry initially and then contaminated with oil afterwards. Something to do with metallic particles within the "Ferodo" friction material scouring the plain steel friction surfaces when dry then becoming contaminated with oil later resulting in a slipping clutch.

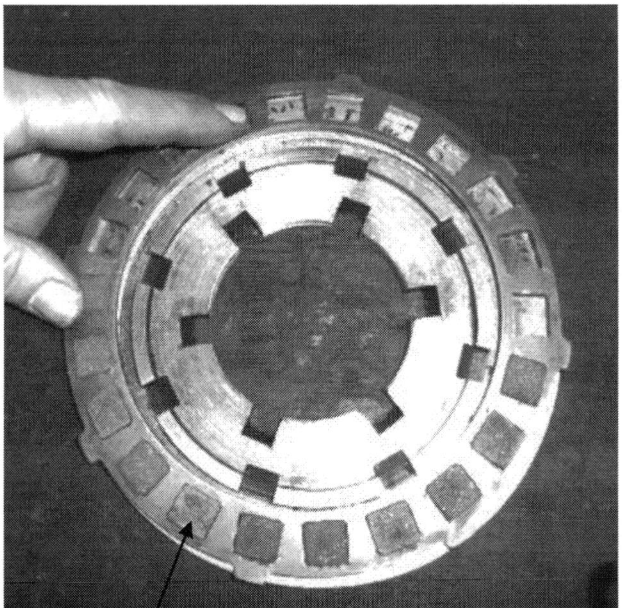

New clutch "Ferodo" inserts

Friction plates submerged in auto gear box oil

The clutch on my KSS is very smooth on take up, does not drag, if un-abused, and disengages easily resulting in smooth gear changes and does not slip.

Clutch Assembly

To build up the clutch assembly it must be said that, if you do have a spare sleeve gear from an old Velocette gearbox, it makes the job that bit easier as it helps line up the plates as you build them. If you do not have a spare sleeve gear, do not despair, as all you need is a good eye, a bit of patience, and a couple of tie wraps.

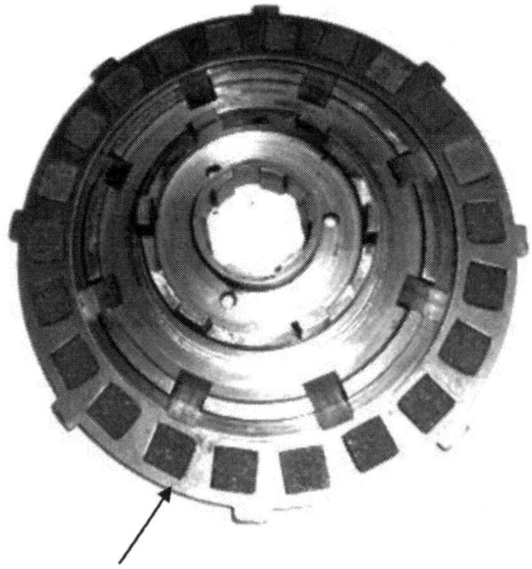

Friction Plate

Start with the back plate and lay it down on a flat surface covered in tissue to keep everything clean. Then place on top a friction plate.

On top of this plate, place a plain plate with the inner tongues facing upwards so that they engage with the slots in the backplate.

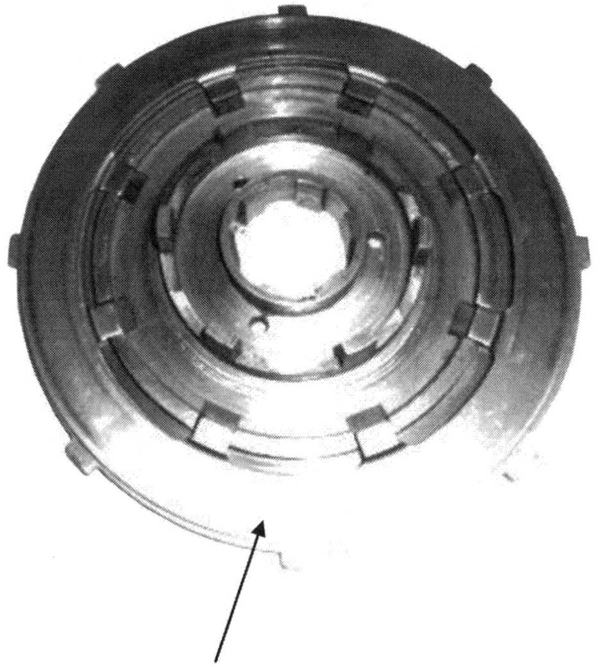

Plain clutch plate

Next, place the centre friction plate which carries the sprocket. Care is required to ensure that the friction plate below outer tongues engage with the slots in the centre plate.

Centre friction plate with sprocket

Ensure friction plates engage with slots in centre plate

Continue with the remaining plates as described until the clutch is fully assembled.

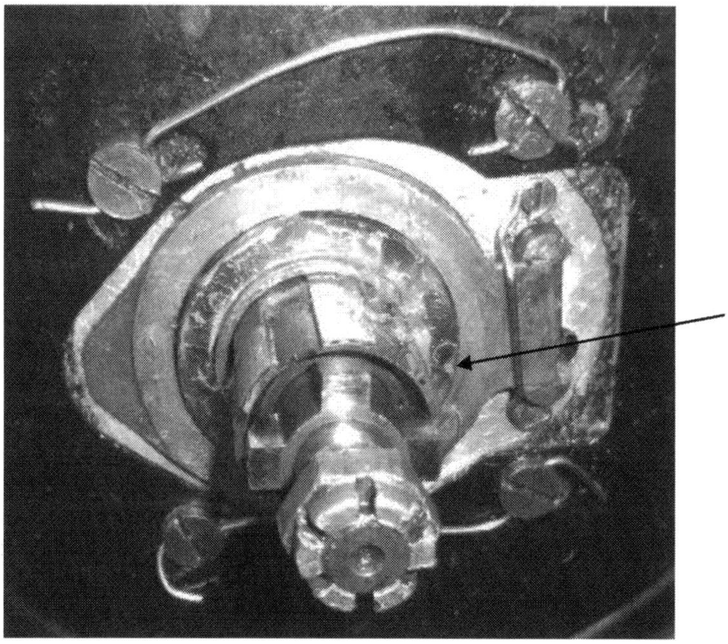

Spherical thrust face clutch release bearing

Note that parallel roller bearings and a modified clutch release bearing cage are recognized as being an improvement over the original ball bearing arrangement.

Finished assembled clutch ready for fitting

In order to keep the clutch assembled without the plates coming apart before the assembly is attached to the gearbox shaft, wrap the clutch together with tie wraps as shown in the photo. Even with two pairs of hands, the clutch, primary chain and engine sprocket take some negotiating to get them both on their respective shafts and tease the chain over the footpeg bracket.

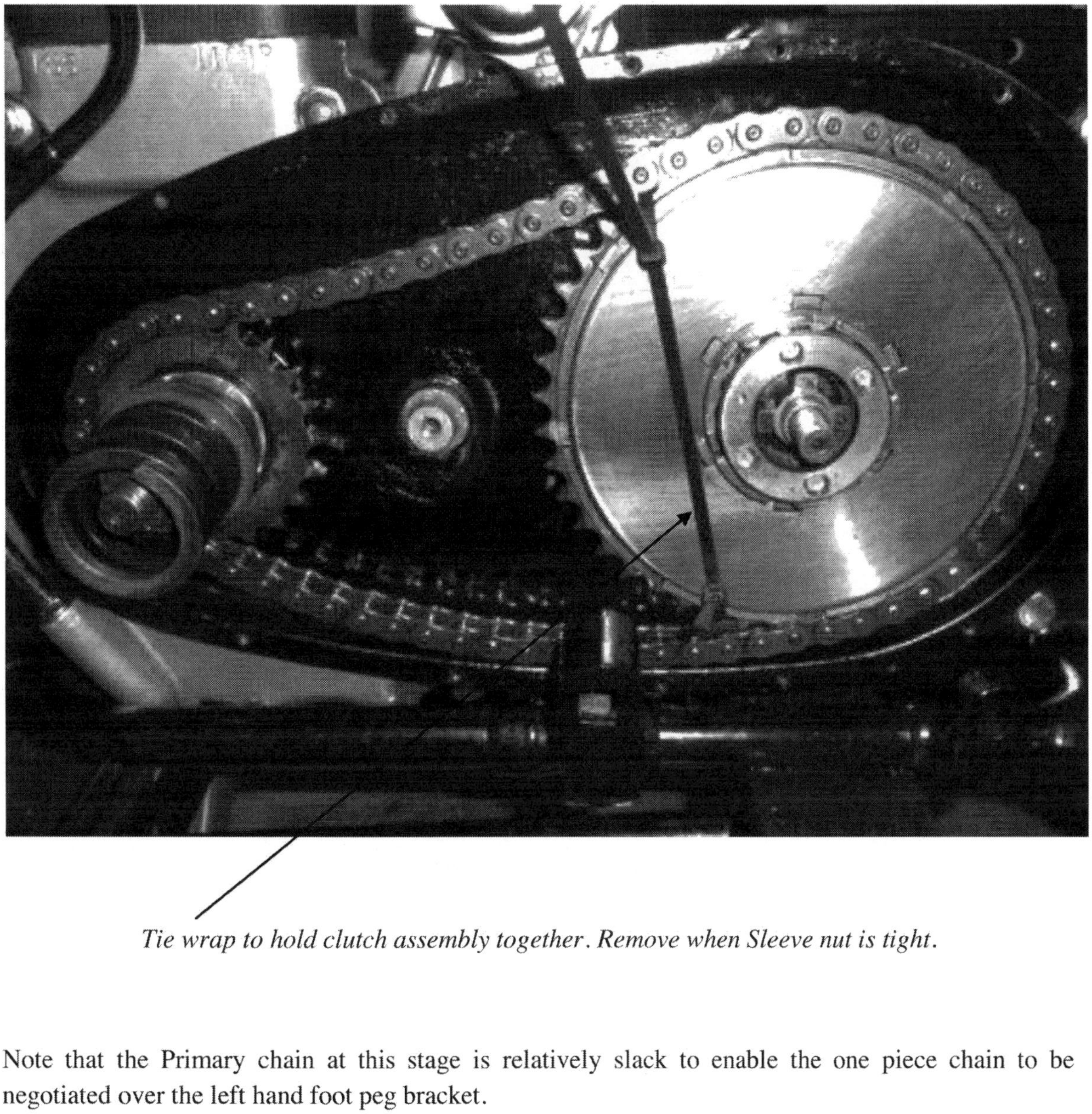

Tie wrap to hold clutch assembly together. Remove when Sleeve nut is tight.

Note that the Primary chain at this stage is relatively slack to enable the one piece chain to be negotiated over the left hand foot peg bracket.

Adjust primary chain to give 12 mm of total slack on the top run by moving the gearbox rearwards.

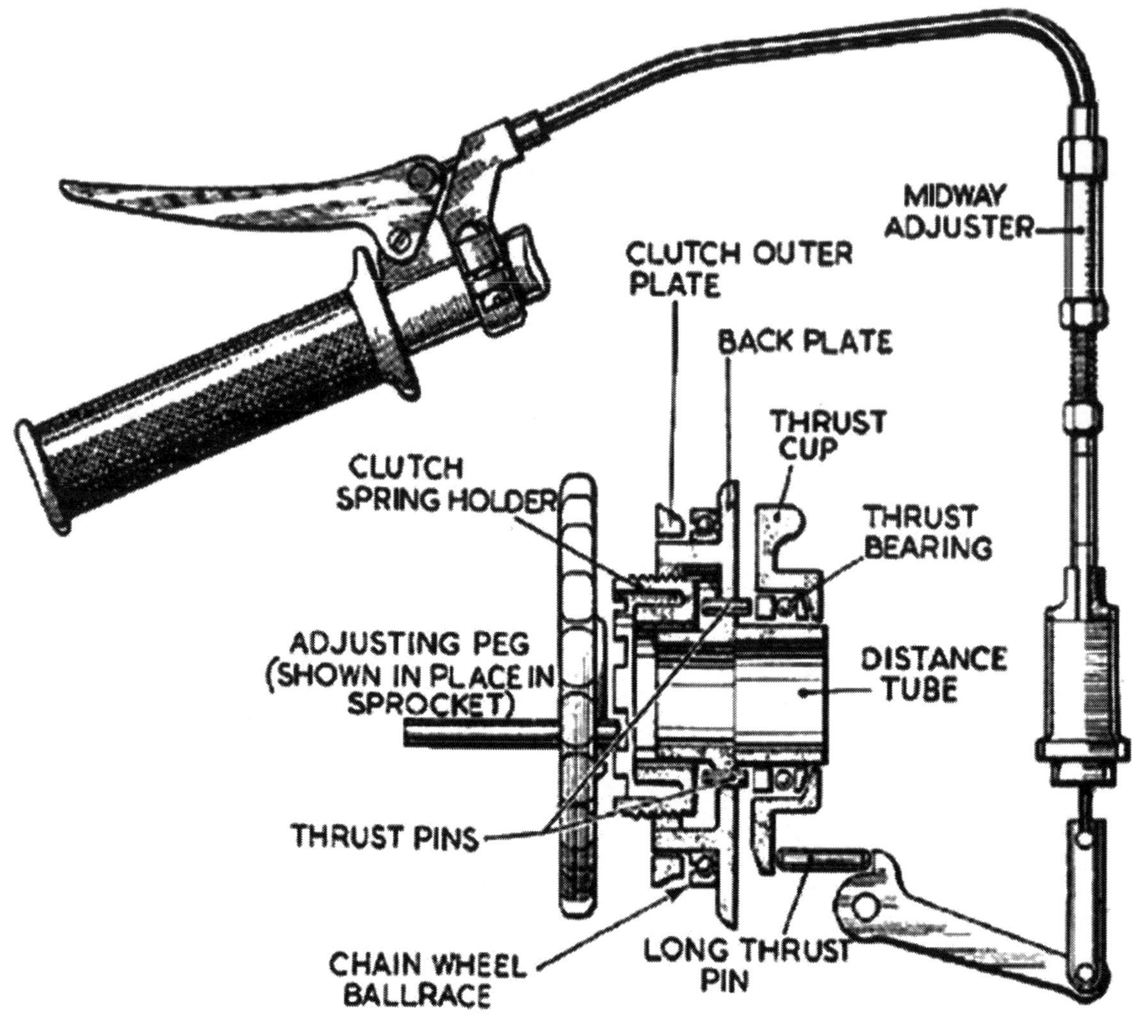

Pictorial representation of clutch operating mechanism

Before assembling the clutch to the gearbox shaft, assemble the spacer that fits between the gearbox output bearing and the clutch backplate. Smear the thrust bearing spherical face, rollers, brass cage, front thrust piece, 3 thrust pins and the rear of the clutch with grease. The grease will hold the three thrust pins in the clutch backplate. Screw in the clutch spring holder into the clutch backplate, about half way, followed with the spring spacer plate with the 16 springs rotated so that the tails towards the outside edge point backwards so that the springs do not bend backwards on tightening the sleeve nut.

Screw the sleeve nut onto the gearbox input sleeve shaft with two new shim washers between the sleeve nut contact face and the springs. Ensure that the sleeve shaft stays forward on the shaft in order to reach it with the sleeve nut. Using either the peg spanner or preferably the four pin tube spanner with T bar, tighten up until the clutch assembly is retained onto the sleeve shaft.

It is at this stage that the clutch release mechanism is initially adjusted to ensure that the thrust mechanism is as close to the gearbox as possible. The thrust pin in the gearbox should be level with the gearbox outer face as at this stage the clutch cable should not be attached to the release pivoted lever pull rod in the gearbox. Screw in the clutch spring adjustment ring until it just makes contact with the three thrust pins. It is at this stage that the sleeve nut can be tightened down fully and torqued up tight.

Tighten the sleeve nut completely up to the clutch rear plate, tight enough to make sure the clutch assembly does not rattle on the sleeve shaft causing damage. In order to provide adequate locking of the assembly to fully tighten the sleeve nut, sprag the engine and clutch sprockets with an aluminium strip as shown in the photograph for disassembly.

Attach the clutch cable to the actuator rod in the top of the gearbox and screw in the spacer tube complete with sealing washer and thimble. Place the rubber shroud over the cable and spacer tube to prevent water from entering the gearbox down the clutch cable. Do not attach the clutch cable to the handlebar lever at this time.

Clutch Adjustment

To provide clutch adjustment, raise the back wheel off the ground on the rear stand. Remove the two small screws holding the circular access cover to the rear chain guard and insert the clutch adjustment tool (a ¼" diameter steel rod with a flat on one end that enters the sprocket first) into one of the holes in the sprocket until it engages with one of the slots in the clutch spring carrier. With the gearbox in neutral, turn the rear wheel backwards a little at a time until the clutch just starts to slip when using the kickstarter with the throttle open. Connect the clutch cable to the handlebar lever and adjust the cable adjuster to remove all cable slack. Tighten the cable adjuster locknut. Then turn the rear wheel forward, a little at a time, thereby turning the clutch spring carrier anticlockwise, until the clutch cable has a 1/8" slack at the handlebar lever.

If the clutch ever starts to slip in service, do not make adjustments to the cable at this time. Follow the procedure above but turn the rear wheel forwards until the clutch stops slipping and the 1/8" cable slack is returned.

If clutch adjustment is totally lost, either due to mal adjustment or relined clutch bedding in, then follow the procedure above. If the clutch still retains its original fault and cannot be adjusted by the above procedures then further investigations will be required to ascertain if the fault is caused by any other reason. i.e. worn out clutch linings or badly worn/distorted/broken components.

If the clutch is fitted and adjusted correctly then experience has shown that the Velocette clutch can provide many thousands of miles trouble free service.

Rear Drive Train

Rear chain sprocket shown attached to brake drum

The rear sprocket on the Velocette KSS is spigoted on to the rear brake drum held on by six $^5/_{16}$" bolts, nuts and washers, the rear brake drum in turn is itself spigoted on to the rear wheel through a three stud drive system by three domed nuts fastening the drum to the wheel. This system, with the use of spacers, a dummy shaft and the main wheel spindle provides for a quick release (QR) rear wheel. This, in effect allows the rear wheel to be removed without disturbing the rear chain drive or the rear brake mechanism.

Quick Release (QR) rear wheel showing chromed domed nuts holding wheel to brake drum

Having replaced the rear brake drum on the KSS for a new one (the old one was badly scored), it became apparent that the $^5/_{16}$" BSF nuts and bolts original configuration would not be feasible. The wall thickness of the new brake drum was a tad thicker than the original so hexagon headed bolts would not fit. I utilised HTS socket headed cap screws fitted from the brake drum side and then fitted thinned nuts to the sprocket side, locked with stud lock. Full width nuts hit the flange on the brake plate hence the modification.

New rear brake drum on left replacing scored old drum

Rear wheel showing chain adjuster and lock nut in the frame
(Note domed nuts holding sprocket to brake drum)

The rear sprocket was replaced with a new one as was the gear box sprocket and final drive chain. These were badly worn as would be expected from such aging components open to all weather and road condition.

Do not finally set the rear chain tension until the primary chain tension has been set. By adjusting the gearbox to take up any slack in the primary chain will in turn reduce the centre distance of the final drive sprockets. It is essential also to ensure that the rear wheel is set straight in line with the front wheel with the steering set to dead straight. String or straight wooden planks can be used for this purpose making adjustments for the different width of tyres from front to rear. Once this is set accurately it is only required to set the adjusters on each side of the wheel spindle by the same number of turns when making subsequent adjustments.

Modern chains have O ring seals at the ends of each roller in the chain that keep the lubricant in and dirt out. That does not mean that the rest of the chain will not go rusty. In the old days we used to boil the chain in graphite grease after cleaning it in paraffin. That got rid of most of the dirt but it was essential to lubricate the chain straight away otherwise it would just go rusty and be ruined. Chain maintenance has come on a long way since then and spray on lubricants are a must with modern chains. The Velocette has a built in chain lubrication system in the vent pipe from the oil tank. If this is supplying excessive amounts of engine oil onto the rear chain and hence onto the rear tyre, then the engine is in need of some attention.

Brakes

The front brake on older Velocette single cylinder machines is a 7 inch, single sided, single leading shoe unit integral with the front wheel hub. The early girder fork front brake also housed the speedo drive mechanism within a steel brake backplate. On the Dowty telescopic fork model as fitted to the very last few KSS Mk 2's, the brake back plate is an aluminium casting, dimensions of the brakes being the same as the early model. The speedo drive in this case is taken off the back wheel.

My KSS front brakes were well worn even though the drum was still quite serviceable. The riveted brake linings were removed and the shoes sent away to a brake specialist to have new linings bonded on to the shoes. New brake shoe retraction springs were fitted to replace the suspect old ones.

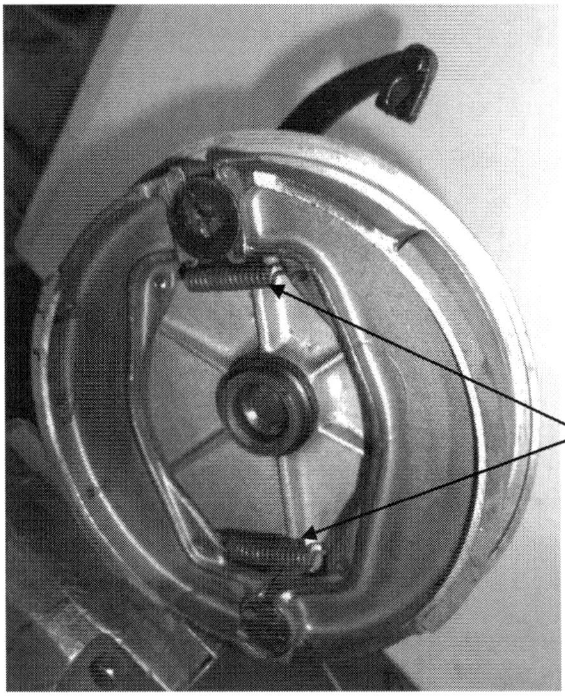

Front brake with relined bonded linings and new release springs

Rear brake with relined bonded linings and new release springs.

Note the new rear drum and sprocket behind.

On reassembly do not forget the two shims between the operating cam and the shoes.

Tanks

Petrol and Oil Tanks

The petrol and oil tanks on all single cylinder Velocettes are substantial sheet steel pressings and fabrications, welded and finished to give a long life. Internal corrosion may cause problems as is often reported when water has been allowed to get in the petrol and the bike left in storage in this state. Luckily the petrol tank on my KSS had suffered only minor corrosion both internal and external, having been dry stored for many years. The oil tank was in pretty good condition apart from a small dent which was easily attended to with a small addition of filler smoothed off to match the contour of the tank. The tank internals were washed out with petrol of the oil residue that had saved the tank from any possible corrosion.

The petrol tank still had a large proportion of the original factory paint finish complete with pin stripes and Velocette tank logos which came in very useful for the final finished logos. Surface corrosion was well evident along with one or two minor dents. The petrol tank was photographed from every angle as a record of the original factory pin stripe lining and logos. Only after this stage was paint stripper applied, taking the tank back to bare steel. Wire brushing and reducing grades of emery paper removed most of the surface rust. The remaining rust pockets were then treated with commercial rust treatments to isolate and inhibit any further corrosion. The minor internal corrosion was mainly removed with small pebble grit, bounced around the tank by tumbling by hand. Good exercise for anyone. The tank was then washed out with high pressure water to remove any residues, and dried with a hot hair gun and a 2 kw electric heater. The tank internals were then treated with commercial rust treatments to isolate and inhibit any further corrosion. A further blast with high pressure water and drying as before. The tank was given a swill of petrol with a small amount of oil to give a surface coating to help prevent any further corrosion until the tank was to be put back into service. The petrol tap holes were plugged with stainless steel pipe plugs sealed with PTFE tape.

Painting was entrusted to a local motorcycle paint specialist who photographed the process at every stage. The tanks were rubbed down to ensure that no residues remained that could affect the later paint processes. Thinners was applied to clean the surfaces then the subsequent processes would not be contaminated. Thin layers of filler were applied over surface imperfections including the minor dents then smoothed off to generate the smooth contours required. All traces of dust were then removed and paint primer finish applied by spray gun. This was flatted down to give a key and finish required for the finishing coats. It was at this stage that the pin stripe preparation was applied. The contours of the pin stripe were roughly masked off by masking tape much wider than the finished stripes. The photographs came in useful here. The gold paint was then sprayed between the two masked areas. When dried, the masking tape was removed and the affected areas rubbed down as before.

To give a perfect width and application of the pin stripes, a self adhesive plastic strip, the same width as the finished pin stripe was stuck over the gold paint precisely, the whole tank was then finish sprayed in gloss black. After giving adequate time to dry and harden, the plastic strips were removed to reveal perfect pin stripes. The whole tank was then rubbed down to remove all imperfections and pin stripe raised edges.

Next, the Velocette logo self adhesive stencils were applied. Note here I say stencils, not transfers or appliqués.

I had purchased Velocette petrol tank appliqués from a "specialist" but the Paint Sprayer pointed out that they did not quite look right compared to the originals on my early photographs. On conducting research on the internet, we found that the definition of the lettering had been lost.

The painting specialist then consulted with a computer logo specialist who came up with the self adhesive stencils. With the aid of the photos I had taken of the tank before the paint was removed, the stencils were positioned in as near as possible in the original positions. See photograph below.

The Velocette logos were sprayed onto the tank using an air brush with the same paint that was used for the pin stripes. This ensured a perfect match, much to my satisfaction and the admiration of everyone who has seen this tank.

The finishing touch in obtaining the high gloss finish was then to flat off the whole tank including pin stripes and the Velocette logos, cleaning off all residues and then spraying the whole tank in clear lacquer. When this coat had dried and hardened, the finish was then enhanced with abrasive polish and a mop. Subsequently, the tanks and all the paintwork have been given plenty of coats of high quality wax polish to enhance the sheen and to give the quality finish a protective coating to help prevent stone chips and scratches.

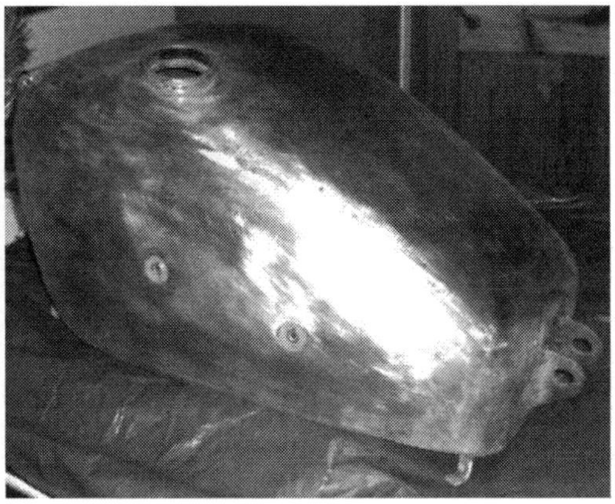

The petrol tank "as found"

All paint and rust removed

Filler applied and flatted off

Tank in grey primer

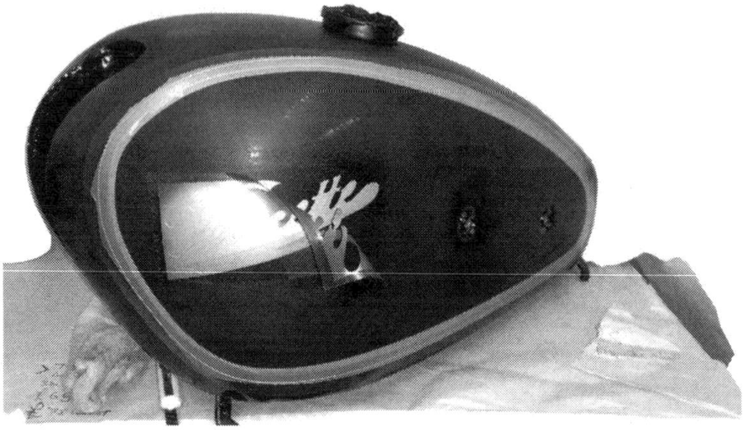

Tank pin striping and Velocette logo stencil

Finished lacquered tank

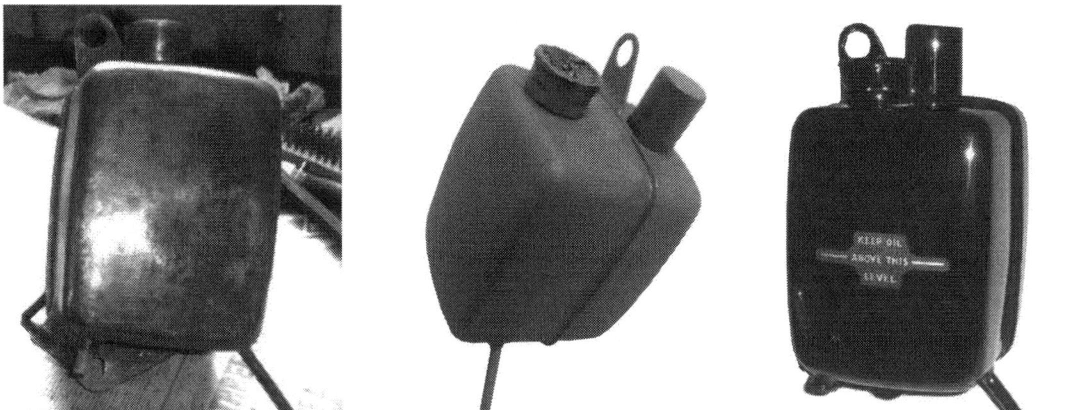

Oil tank at various stages of preparation and final finish with "Oil Level" transfer added.

Controls

Throttle and Levers

The control levers were all original but badly tarnished and the chrome plating was practically worn off the brake and clutch levers. Chrome plating was undertaken by a specialist in Manchester as the original levers were all made of solid brass which is much more in keeping with the motorcycle than the modern replacements made from pressed steel. Threads and screws in the levers were repaired or replaced as necessary. The chrome plating was all done together in order to get a good job done at a good price.

Some restorers prefer to leave the original brass levers unplated and kept highly polished and/or lacquered to preserve the sheen. Personal taste dictates the finish to adopt. My preference was for chrome plate to match the bike as a whole and to retain originality.

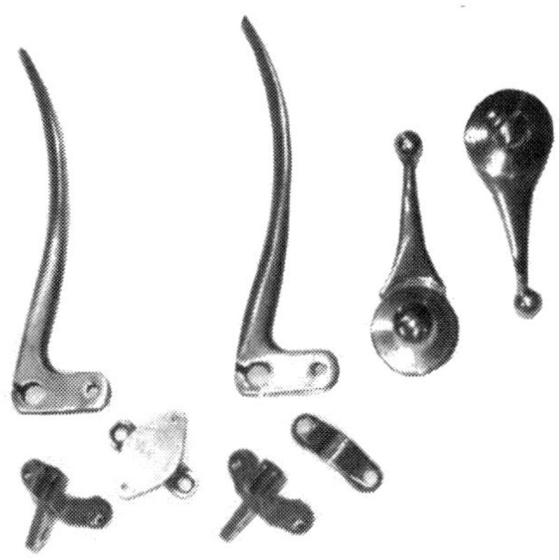

Original brake, clutch and control levers

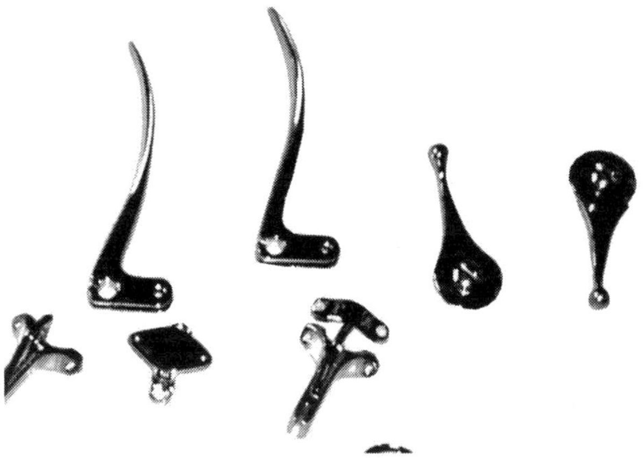

Finished levers in high quality chrome plating

Throttle twist grip body

with fresh chrome

Electrical Systems

All of the electrical equipment on the KSS Velocette, bar the magneto, are Miller which were used extensively on a lot of pre and post war British motorcycles. Later models used Lucas equipment as more machines went over to alternator generators and coil ignition for better lighting and more dependable spark generation using suppressed systems to prevent radio and TV interference.

Magneto ignition systems do not function well on resistance suppressor plugs and caps but, on the plus side they work well on unsuppressed systems and do not depend on a battery.

Dynamo

The Miller dynamo is a compact, well designed electricity generator with a built in cut-out, the voltage control unit being held within a small metal box on top of the dynamo. Later single cylinder models had this voltage control unit positioned under the saddle.

Miller dynamo spare parts are difficult to obtain apart from coils, armatures, bearings and brushes. That being said, these are the most common items required but the replacement armatures are expensive. The cut-out and voltage control unit are no longer manufactured as it is easier to use modern electronic components that can be used, hidden within the confines of the dynamo end casing and the original voltage control metal box.

The dynamo is relatively easy to strip, clean and re-bearing. New brushes are simple to install and the armature can be renovated by skimming the armature in a lathe. The insulated segments in the armature require undercutting by 0.025" to ensure no voltage leakage across windings.

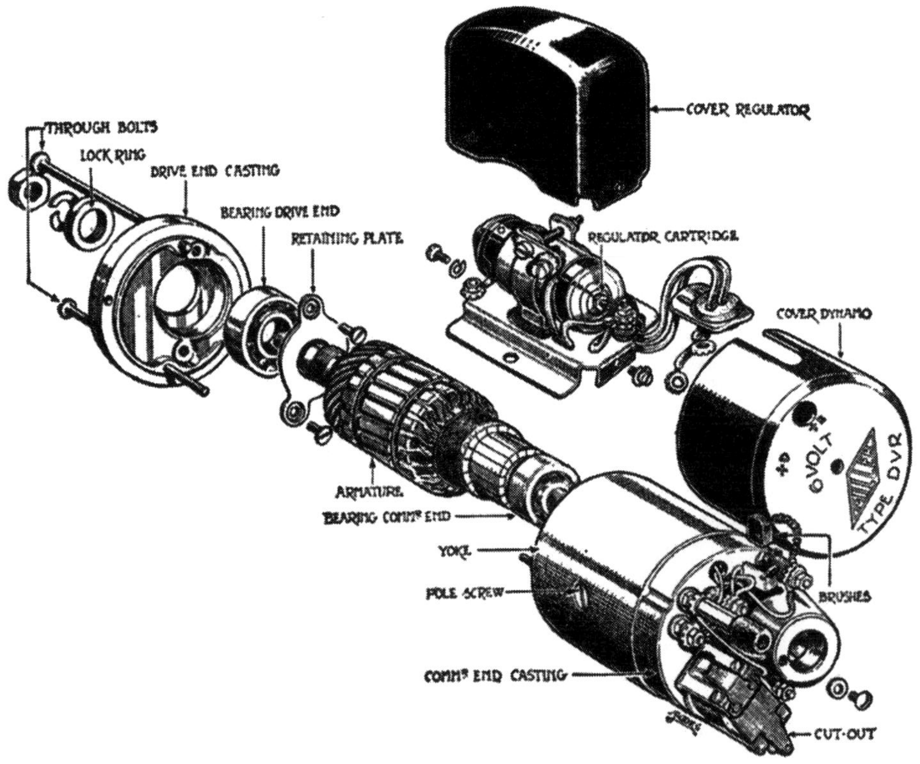

Exploded View of the Miller DVR Dynamo

A new field coil replaced the old suspect coil and the dynamo was converted to operate as a Lucas unit by removing all the brass strips behind the backplate along with the cut out as this was no longer required with the incorporation of the electronic voltage regulator. The dynamo relied purely on hard wiring to two terminals, i.e. F & D. These were connected to the electronic voltage regulator as shown in the photograph and illustrated in my wiring diagram at the end of this book.

Dynamo ready for stripdown

Armature showing wear on the commutator

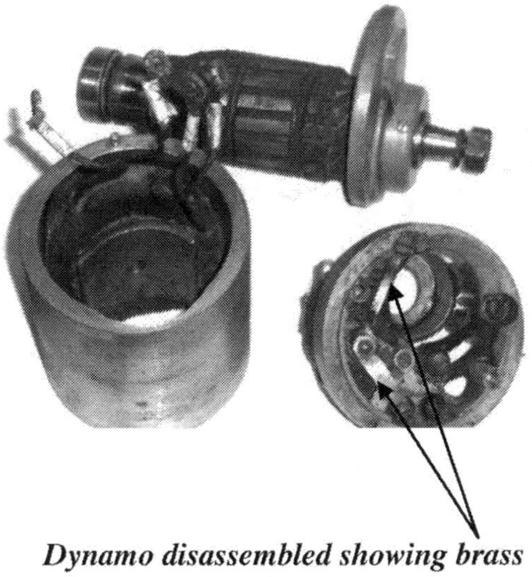

Dynamo disassembled showing brass links behind the front cover

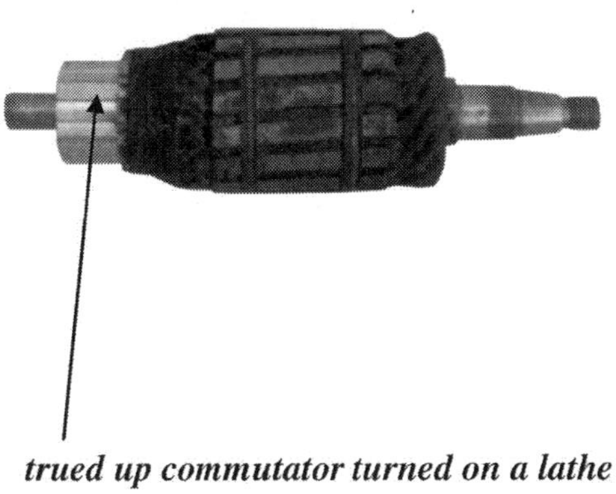

trued up commutator turned on a lathe

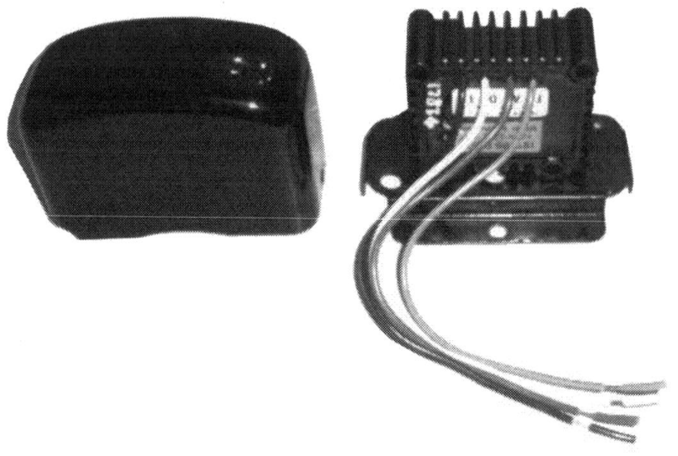

Electronic Voltage Regulator mounted in the old VR case

All connected up and ready to go

Dynamo & VR wiring sealed and wrapped with end cover fitted

Magneto

The magneto as fitted to the Mk 2 KSS Velocette motorcycle is the BTH Type K.C.1-L4.

BTH Magnetos rarely need any maintenance over long periods of time, other than to check that the points are adjusted to the correct gap and the cam is lubricated sparsely with thin oil ensuring no oil gets onto the points.

I recommend that any major failures of a magneto resulting in a weak spark or no spark at all, the magneto be entrusted to a specialist for refurbishment. Specialist equipment is required for disassembly and testing purposes which most private individuals would not have access to.

After a long period of time, not forgetting that most of these machines are well over 70 years old, the bearings, windings, capacitor and indeed the permanent magnets will all be suspect. My magneto worked a bit but would not give satisfactory starting and slow running without putting a new spark plug in the engine every time. After a complete refurbishment involving new bearings, a rewind, new capacitor and degause followed by a re magnetization it transformed the engine into an easily started and very good slow running machine.

BTH Magneto refurbished, rewound, new capacitor, degauzed and re-magnetised

Points all cleaned up and reset to a gap of 0.012"

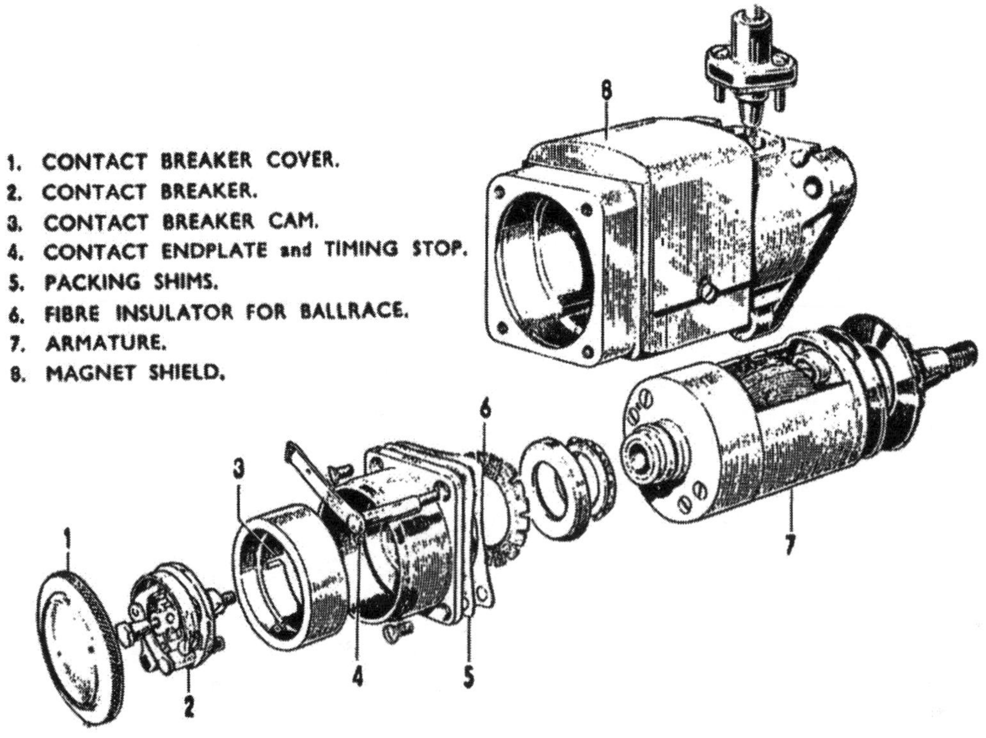

Exploded view of the BTH magneto as fitted to a KSS

Wiring Loom

Complete wiring loom as purchased

As befitting all basic vehicle restorations, a new wiring loom makes for a better looking and reliable electrical installation. I purchased a complete loom recommended for the KSS Mk2 but found that the insulation colours do not match the original. This is not surprising as the new British Standard colour schemes for conductors on motor vehicles follow a similar code so that electrical functions may be identified from the colour combinations. e.g. Brown for main power feeds. green for front and rear pilot lights, and then blue for head lights.

This is where complexity in restoration makes a diversion in that we seem to have variations on a theme when other components are used such as a headlight dipper switch. The cables for main and dip beam on my switch are green and yellow so I have recognised this along with other variations in my new wiring diagram later on in the book. On vehicles with more complex electrical systems, more colours are used.

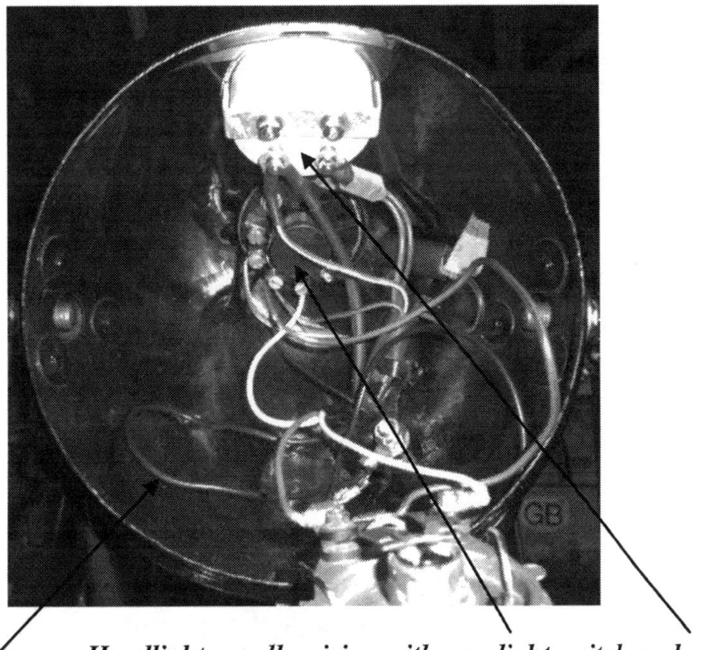

Headlight nacelle wiring with new light switch and ammeter.
(Note the black wire grounded earth connected through loom to frame earth)

Pre Running Procedures, Priming of Oil System. Oil Pressure Setting and Determination of Lub Oil Flow Rates

The lubrication of the engine is at the heart of efficient, effective and reliable running. When starting an engine for the first time from a strip down or rebuild, it is essential to ensure that everything is functioning correctly before taking the bike out on to the road. To achieve this certain things need to be in place beforehand.

1. Ensure there is adequate clean oil in the tank of the correct grade.
2. The oil filter in the tank is installed and undamaged.
3. Piping to and from the engine is intact, held on firmly at each end with appropriate pipe clips.
4. Anti sumping valve, if fitted, is turned on.
5. Pressure relief valve balls and springs are fitted correctly.
6. The oil inlet pipe and pump are primed with engine oil prior to connecting to oil tank.

Before attempting to start the engine for the first time after a rebuild, and assuming all is correct as above, remove the spark plug and with the kick starter, turn the engine over quite a few times to get the oil flowing. A squirt of engine oil through the plug hole with an oil can will ensure that the rings and cylinder bore do not run dry. Repeat this for as long as necessary until oil returns from the stand pipe in the oil tank, just visible below the oil tank filler neck. The oil at ambient temperature will only return in spurts until the engine gets hot when the oil will then return in a steady stream.

When the oil return has been confirmed, screw back in the spark plug, connect the HT lead and prepare for starting. Start the engine as described in the next chapter. When the engine has reached running temperature and the oil in the tank is quite warm, set the engine speed to fast idle. Remove the oil filler cap and check the oil is returning in a steady stream.

At this stage it will not be known if oil is getting to the cam box and the only way to find out if it is, is to look at and measure the oil drain from the cam box to the crankcase. This is done by disconnecting the drain line at the bottom of the Y piece on the left hand side of the engine and holding a collection jug underneath. With the engine running at fast tickover, measure the amount of oil returned in one minute. There is no hard and fast rule but general consensus amongst KSS owners is about 0.1 Litre per minute.

When oil return from the engine and the cam box has been confirmed, stop the engine and remove the blanking plug on the top bevel housing cover. Replace this with a $^1/_8$ BSP pipe connecter, a length of plastic piping and a suitable pressure gauge with a range of 0 – 15 psi. Restart the engine and leave running at fast idle. The oil pressure should gradually rise until about 8 – 9 psi has been reached. Increase the engine revs gradually and the oil pressure should rise between 11 -13 psi.

If the oil pressure is inadequate, then check that all lines are free from obstruction and the procedures at the beginning of this chapter are followed. If everything seems correct, then check that the oil pressure adjustment screw at the rear of the right hand crankcase is set correctly and that the ball is seating correctly and that the spring is not damaged. If all appears to be present and correct, adjust the relief valve screw inwards a little at a time and observe if the oil pressure has improved. If not, then matters will need to be further investigated as to the cause of the problem.

Once the flow rates and oil pressures have been confirmed, remove the oil pressure gauge and replace the blanking plug with sealing washer back into the bevel drive cover.

Oil pressure gauge with upper bevel housing tapping blanking plug

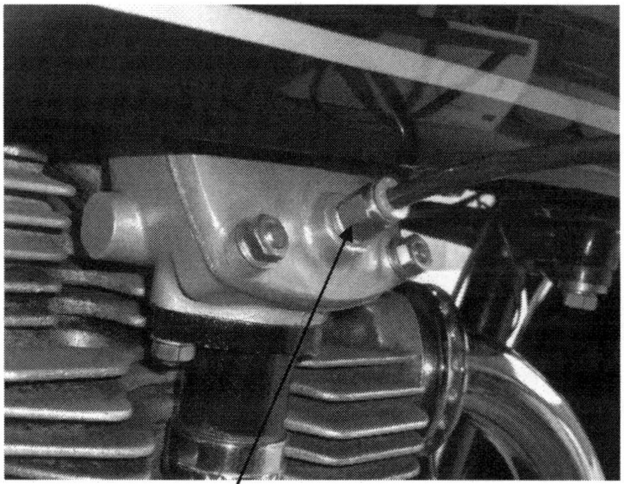

$^{1}/_{8}$ BSP Pressure hose connection to remote pressure gauge

Top bevel housing oil pressure = 9 psi at fast tick over. Oil temperature hot normal running.

Scavenge return to tank – normal steady flow.

Cam box oil drain flow rate at fast tickover = 0.1 lt / min

Oil pressure adjustment screw location

The oil pressure adjustment screw is located adjacent to the oil tank connection tapping below the oil pump in the timing chest. Undo the lock nut on the adjustment screw then turn the screw clockwise to increase oil pressure and anticlockwise to reduce. Retighten lock nut after the oil pressure has been confirmed.

Starting the Engine

Starting any model of a single cylinder four stroke Velocette requires a certain knack. In Velocette parlance this is often referred to as "The Drill".

Sounds quaint but I believe this came from the past when people spoke a little different to the way we speak today.

Every Velocette owner will have their own particular method but each essentially follows the same routine.

In order to ensure practically first time starting, it is essential that the engine is in good working order, is regularly serviced, has adequate fuel of the correct grade in the petrol tank and that the owner has developed a fair grasp of the manual settings for the engine and the technique involved. One has to be firm with the kickstarter without being brutish.

The sequence of events for starting a cold engine are as follows:

1. If an anti sumping valve is installed in the oil supply to the engine from the oil tank, ensure it is turned on.
2. Turn the tick over screw lever on the carburettor to the right as far as it will go. This is the starting position.
3. Set the manual retard and advance lever to retard the ignition between $^1/_8$" and a ¼" movement on the lever bezel. This amount will be determined by the rider from experience. Every engine will be different depending on how they have been set up.
4. Set the air lever inwards to bring the air slide down thereby restricting the air flow in the carburetor and hence richening the fuel mixture.
5. Turn on the left hand petrol tap. The right hand tap is the reserve, being set higher in the tank.
6. Tickle the carburettor float chamber by depressing the button on top of the float chamber top cover up and down until petrol starts to appear out of the little hole in the side of the top cover.
7. At this point some riders prefer to tilt the machine to the right which also has the effect of richening the mixture.
8. Pull the valve lifter lever to open the exhaust valve whilst turning the engine over two or three times with the kick starter to introduce fuel into the engine.
9. Let the valve lifter go then by using the kick starter turn the engine until compression is reached.
10. Bring the kick start lever right up to the top and by pulling the valve lifter, depress the kick starter slowly down to the bottom of its stroke to bring the engine just over top dead centre.
11. Release the valve lifter, and bring the kick start lever to the top position.
12. Ensuring that the kick starter ratchet is engaged, give the kick start lever a good sound swing causing the engine to rev freely over the compression stroke and hence spring the engine into life.

13. Immediately pull the air mixture lever back to raise the mixture slide in the carburettor which will make the engine run with a steady beat and turn the tick over lever to the left on the carburettor to reduce the tick over speed. This may have to be done in phases until the engine gets warm to reduce the risk of stalling.
14. If the engine does not fire, or kicks back punching your knee up into the air, the ignition is too advanced. Keep the knee bent always in anticipation of a kick back then no harm will be done.
15. Cuss and swear at this point, retard the ignition a little and try again.
16. Practice makes perfect and there is no substitute for getting your technique right that suits yourself and the engine.
17. It is said that many a prospective Velocette owner is put off by the reluctance of the engine to start and this can be put down to the rider not following the basic technique.
18. If "The Drill" is perfected then first time starting can more often than not be guaranteed.

Many a joke is made about the Velocette's reluctance to start, my favourite being the guy who said that if he did not live on the top of a hill he would never have bought a Velocette.

I think that broken bones and dislocated knees due to the kick back on starting a Velocette are things of legend but I remember a funny incident many years ago when I was a teenager.

A friend of mine had a beautiful 500cc Venom Clubmans which was only 3 or 4 years old at the time. He was a strong lad and usually had no trouble starting the bike. He had just fitted a full Avon Dolphin fairing with the headlight behind a Perspex nose cone. He had just replaced the standard single curvature screen with a double curvature screen at great expense. Having fitted "clip on" handle bars, his reach with his arms brought his right elbow in line with his right knee. You have guessed it. On trying to start the engine, it kicked back, throwing his right knee into his elbow followed by thrusting his arm into the expensive double curvature screen and breaking the screen. The air was blue that evening.

Tools

One can strip a Cammy Velo without the following tools but it is not recommended if you want quality of workmanship to be your aim.

Used in conjunction with and supplemented by a good range of Chrome Vanadium sockets and combination ring spanners covering all sizes required on the machine in British Standard Whitworth (BSW), British Association (BA), Unified AF (If you have seccumbed to using modern day nuts and bolts). Socket sets in ½", $^3/_8$" and ¼" square drive are invaluable. Square ended flat screwdrivers of the correct size for the screw head will save mal-forming screw head slots and embarrassing scratches on covers and castings.

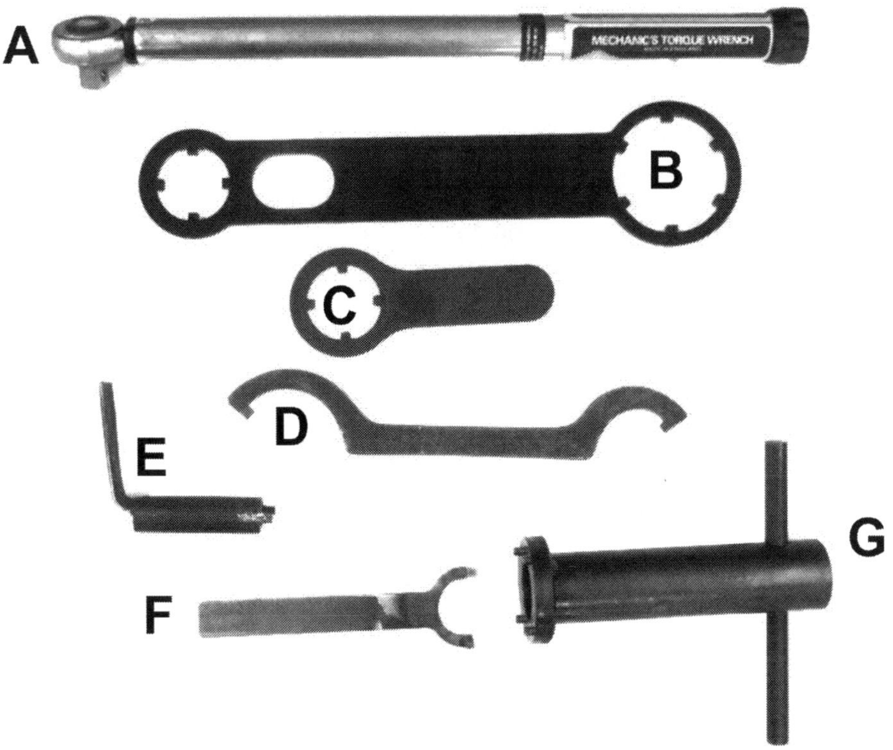

A – Torque Wrench
B – Crankshaft Nut / Clutch Spring Carrier Holder
C – Crankshaft Nut LW version
D – "C" Spanner Clutch Spring Carrier Holder LW
E – Slotted tongued wrench for Dowty forks seal carrier
F- Cranked tongue "C" spanner for cam drive gland nuts
G – 4 pin box wrench with T bar for Clutch Sleeve nut

Spares and Services Suppliers

Sources of spares for all Velocette motorcycles are quite good considering the age of the vehicles, some going back into the earliest decades of the 1900's. Spares for the KSS range are readily available, the only components difficult to obtain being crankcases, barrels and cylinder heads.

Pistons, rings, bearings, valves, guides and gaskets are always available. Some sources are better than others but I have found that it pays to shop around for critical items such as pistons and bearings.

Petrol tanks and tinware can be bought readily off the internet but again be careful of quality.

The main sources of spares and services I have used are listed below.

Velocette Owners Club Spares Dept. Veloce Spares Ltd, The Old Chapel, Cheney End, Huncote, Leicester. LE9 3AD. VSL website www.velocespares.co.uk

Grove Classic Motorcycles Limited, website www.groveclassicmotorcycles.co.uk

Burlen Limited (for AMAL carburetors) Spitfire House, Castle Rd, Salisbury, Wiltshire, SP1 3SA. website: www.burlen.co.uk

Vintage Tyre Supplies Ltd, The National Motor Museum, Beaulieu, Hampshire, SO42 7ZN. website: www.vintagetyres.com

Westwood Cylinder Liners Ltd, Unit 3, George Baylis Court, George Baylis Road, Droitwich, Worcs, WR9 9RB. Website: www.westwoodcylinderliners.co.uk

Newman Cams, Farnborough Way, Farnborough, Kent, BR6 7DH.
Website: www.newman-cams.com

HB Bearings, Riverside Works, Horley, Holmfirth, Huddersfield, HD9 6PQ
website: www.hb-bearings.com

Alpha Bearings, 36 Kingsley Street, Netherton, Dudley, West Midlands, DY2 0PZ
website: www.alpha-bearings.com

Bannister Engines, Auto Component Machining Specialist, Init 1A, Hecla Works, Mounsey Road, Bamber Bridge, Preston

Chrome Plating. Leigh Metal Polishers, Swinton Ind Est, Off Pendlbury Rd, Swinton, Manchester. M27 4AG.

Paint Spraying. RPM Racepaints, Leyprint Mill, Leyland Lane, Leyland, Preston. Lancashire. PR25 1UT. Contact: Ray. P, Mayers, Telephone: 07970138897

Technical Data

Engine

Capacity:	348cc
Bore:	74mm
Stroke:	81mm
Spec:	Two valves, single bevel driven overhead cam
Output max:	20 BHP (thereabouts) at 6,000 rpm
Torque max:	adequate (in old Rolls Royce parlance)

Conrod roller bearing big end

Conrod plain bearing little end

Drive side main bearing double row roller race

Timing side main bearing ball race

Crankcase and cylinder head material, aluminium

Cylinder Material, cast iron with spun cast iron pressed in flanged liner

Inlet Valve material, High Chrome Steel

Exhaust Valve material, Heat Resistant High Chrome Steel

Valve Guides, original, - bronze. New, - sintered, compatible with unleaded fuel.

Gearbox

Four speeds. One up, three down, foot change.

Constant mesh, dog tooth gears, upper main shaft, lower idler shaft.

Brass gear change forks, peg operated from positive stop cam plate, operated via external levers and rod to adjustable position gear lever on top of gear box.

Gear box ratios:

Top:	6.02 : 1
Third:	7.26 : 1
Second:	9.55 : 1
First:	13.8 : 1

Neutral between first and second gear.

Cam no.	Inlet Opens Before T.D.C.	Inlet Closes After B.D.C.	Exhaust Opens Before B.D.C.	Exhaust Closes After T.D.C.	Clearances for Checking Timing		Running Clearances		Ignition Setting Full Advance Before T.D.C.	Model
					In :	Ex :	In :	Ex :		
K 17	37°	52°	65°	36°	.012"	.012"	.010"	.015"	42°	Early O.H.C. Obsolete
K 17-2	40°	56°	68°	47°	.012"	.012"	.010"	.015"	42°	O.H.C. to end of 1931
K 17-3	38°	60°	62°	40°	.012"	.012"	.012"	.020"	35°	DT Engine
K 17-4	43°	70°	68°	48°	.012"	.012"	.012"	.022"	42°	KTT 1931
K 17-5	39°	69°	60°	40°	.012"	.020"	.012"	.020"	42°	KSS and KTS 1932/35
K 17-6	35°	74°	71°	31°	.020"	.020"	.015"	.025"	35°	KTT Mk. IV Obsolete
K 17-6	51°	57°	71°	43°	.020"	.020"	.015"	.025"	35°	KTT Mk. IV and Mk. V
K 17-7	35°	65°	70°	30°	.025"	.025"	.006"	.010"	40°	KSS and KTS Mk. 11
K 17-8	55°	65°	75°	45°	.020"	.020"	.015"	.025"	32°	KTT Mk. VI
M 17	50°	60°	70°	40°	.010"	.015"	.003"	.006"	40°	MOV and MAC
M 17-2	50°	60°	70°	40°	.015"	.020"	.003"	.006"	40°	MSS
M 17-3	30°	60°	60°	30°	.025"	.025"	.005"	.010"	40°	MSS
K 17-9	30°	60°	60°	30°	.025"	.025"	.006"	.010"	40°	KSS and KTS Mk. II
K 17-10	34°	47°	64°	29°	.025"	.025"	.006"	.010"	38°	KSS and KTS
M 17-4	50°	60°	70°	40°	.030"	.035"	.005"	.005"	38°	MAC
M 17-5	30°	60°	60°	30°	.030"	.035"	.005"	.005"	38°	MAC
M 17-7	19°	49°	19°	49°	.030"	.030"	.005"	.005"	38°	MAC , MSS
LE	20°	55°	55°	20°	.006"	.006"	.005"	.006"	-	LE 150 and 200

Camshaft and Ignition timing Data for all early Velocette Motorcycles

Clutch

7 inch multiplate (7) 4 plain, 3 friction plates (22 Ferodo segments in each)

Central friction plate integral with primary chain driven sprocket and central ball race bearing ensuring concentricity of clutch and sprocket to main gear box drive shaft.

Number of clutch compression springs: 16

Transmission

Primary Drive:	Roller chain 0.5" x 0.305" x 67 links
Sprocket Sizes:	Engine 21 teeth, Clutch 44 teeth
Secondary Drive:	Roller chain 0.625" x 0.375" x 99 links
Sprocket Sizes:	Gearbox 17 teeth, Rear wheel 46 teeth

Wheels

Front wheel: 21inch WM 1, lattice spoked on twin ball race bearings integral with front brake drum.

Rear wheel: 20 inch WM 2, lattice spoked on twin ball race bearings, quick release 3 domed nuts held to rear brake drum studs mounted on a dummy shaft, via a secondary shaft screwed into dummy shaft, all mounted in lugs at rear of frame.

Tyres

Front: Avon Speedmaster 3.00 – 21R Tubed Type

Rear: Avon SM 3.50 - 20 58H (3.25x20) Tubed Type

Brakes

Front: 7 inch single leading shoe. cam/cable/hand lever operated

Rear: 7 inch single leading shoe, cam/rod/foot lever operated

Electrics

Miller 6 Volt negative earth
10 Amp Miller DVR dynamo with built in cut out and
Automatic voltage regulator unit fitted as standard.

Spark Plug

Champion N5C or

NGK B6ES

Magneto

BTH Type K.C.1-L4

Dynamo

Miller DVR

Voltage Regulator

Miller CVI used in conjunction with Miller cutout in dynamo end casing as standard, or with

NanoControls voltage regulator V-Reg II with Miller dynamo converted to Lucas operation

Wiring Diagram (Miller)

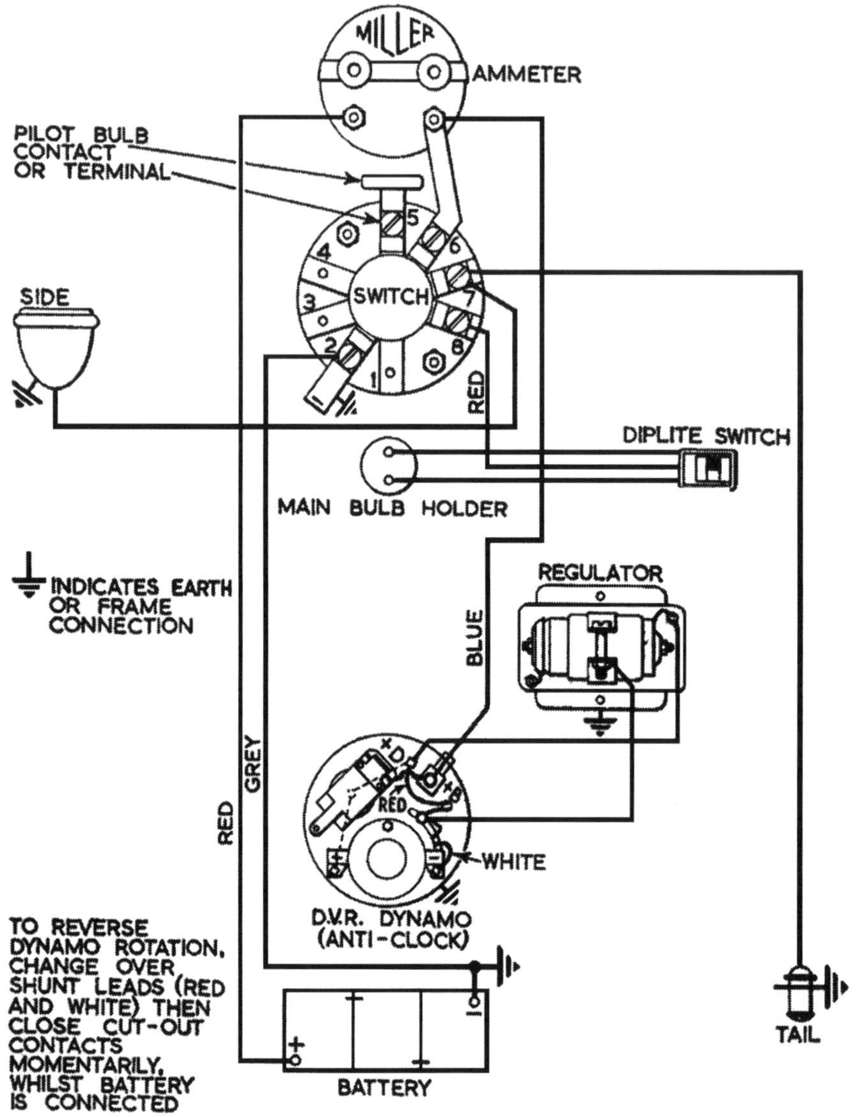

My Velocette World
Snippets of the authors collection of Velocette knowledge

Useful reference books for information on Velocette motorcycles

Book on the Velocette. by Ferrers Leigh, published by the Pitman's Motor Cyclist's Library

Instruction Book for Velocette Mark II Overhead Camshaft 350cc KSS and KTS, price 2/6d nett. An official Velocette publication

Velocette Motor Cycles, A Practical Guide Covering Models from 1933, by R W Burgess. published by C Arthur Pearson Limited.

Velocette Singles 350 – 499 cc 1953 – 1971. Owners Workshop Manual, published by Haynes

KSS Super Profile. by Jeff Clew. A Foulis Motorcycling Book published by Haynes Publishing Group

Velocette An Illustrated Profile of Models 1905 – 1971 by Dave Masters, published by DMD Publishing.

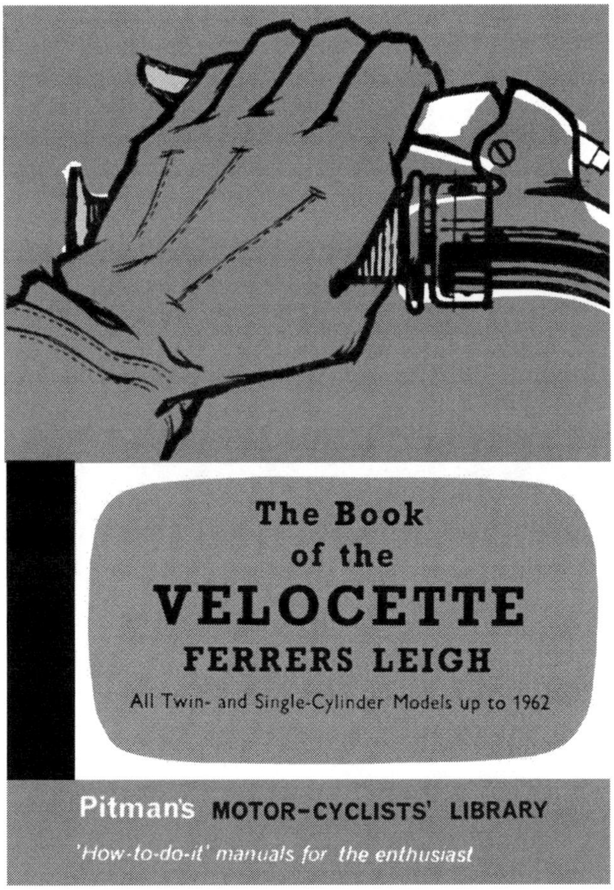

INSTRUCTION BOOK

FOR

Velocette

MARK II OVERHEAD CAMSHAFT MODELS
350cc.

KSS AND KTS

PRICE 2/6 NETT

VELOCE LIMITED
HALL GREEN WORKS
BIRMINGHAM, 28
ENGLAND.

Telephone: SPRingfield 1145/6/7.
Telegrams: "Veloce, Birmingham."

VELOCETTE MOTOR CYCLES

A PRACTICAL GUIDE COVERING MODELS FROM 1933

By
R. W. BURGESS
Veloce Ltd.

With 67 Illustrations

LONDON
C. ARTHUR PEARSON LIMITED
TOWER HOUSE, SOUTHAMPTON STREET
STRAND, W.C.2

VELOCETTE
Singles
349 cc - 499 cc. 1953 to 1971
OWNERS WORKSHOP MANUAL
Haynes
THE BOOK
LDL 154

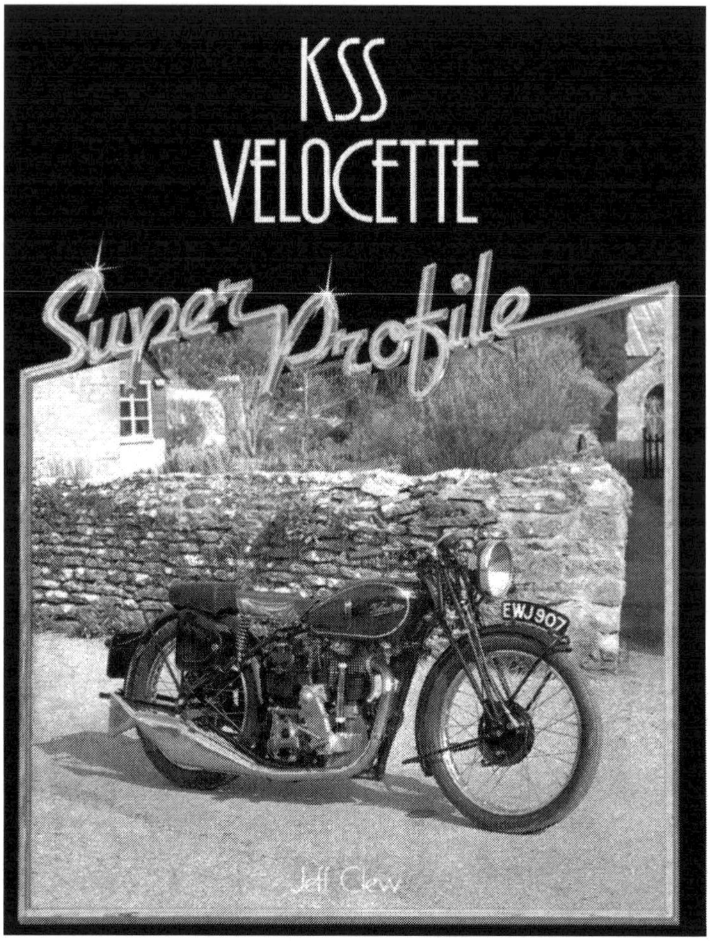
KSS
VELOCETTE
Super Profile
EWJ 907
Jeff Clew

Velocette
Velocette
AN ILLUSTRATED PROFILE OF MODELS
1905 - 1971
DAVE MASTERS

An article published in VOC magazine FISHTAIL
Velocette KSS Mk2 in the Classic Bike of the Year Competition
Held at the Motorcycle Show at the EXCEL Exhibition Centre LONDON
Friday 17 Feb to Sunday 19 Feb 2017. Wrote by the author.

After I sent my application pro forma out of the Classic Bike magazine along with a couple of photos of the KSS taken in my garage, I received a phone call from the magazine wanting to make an appointment with the magazine's photographer. The resulting photos brought the bike alive which looked the part in my untidy garage.

Sub-sequentially, the magazine's Technical Editor questioned me for about 30 minutes on all aspects of the bikes restoration. The main error came in the basic description when he billed the bike as a 1939 model instead of 1947, which I had told him. To the man in the street it probably would not have meant much, but to the Velocette enthusiast it must have rang alarm bells. Being one of the last KSS Mk 2s, it is fitted with Dowty telescopic front forks. So, with that in mind, I thought that my chances had gone with the enthusiast. It did not win but the bike got lots of praise from enthusiasts, specialists and the public. There was only 1 winner and 9 joint second places!

I bought the bike 6 years ago, complete, but in many pieces, after placing a Wanted advert in Fishtail. On getting it home in a trailer, I found that all that was missing was the gear lever. Astounding! Buff log book and V5 logbooks showed matching numbers, confirmed later by Ivan Rhodes from VOC factory records.

Six years later, with a triple heart by-pass in between, the bike was finished. In detail, the rolling chassis was soon put together with a cleaned up and fresh application of glossy black powder coating on the frame. The previous owner had the front and rear wheels fitted with new rims and spokes and the tin ware and tanks were in remarkable condition for the year be it a little rusty. All solved with a good clean up and respray.

Next was the gearbox that came in pieces. The lay-shaft had suffered the usual Velo malady of residing in water with horrendous corrosion being evident. The solution was found with a trip to the Stafford show where I obtained another gearbox from which I transplanted the gears, selectors and cam-plate into the KSS box, along with new bearings, shield and other bits.

The clutch was given the same treatment with the replacement of the thrust bearing for a roller type and fitting new friction inserts (22 in each plate). The primary sprockets were in remarkable condition so a new primary chain was all that was required. The final drive was a different story, notwithstanding being open to the elements it was still serviceable, just! I fitted new sprockets and drive chain along with a new rear brake drum with new brake linings bonded onto the existing shoes with new return springs. The same was done to the front brake but the drum had little scoring being completely serviceable.

The engine was my biggest challenge, it was stripped down systematically to the individual components, cleaned up of all the internal tarish gunge, and then painstakingly assessed as to what was required to bring it back to life. I fitted new main bearings, new big end and little end from Alpha Bearings, adjustments made with shims to get the 0.001" end-float essential for bevel gear quiet running. I re-sleeved the cylinder and re-bored to suit the piston and new piston rings.

I stripped the cylinder head completely and fitted new sintered valve seat inserts supplied by the engine machine specialist who had re-sleeved the barrel. New valves, valve guides and springs made up the assembly. I made new valve stem seals out of leather strip. The cams were a little worn as were the rockers which were all re-profiled by Newman Cams. The cam drive bevel gears were in good condition as was the vertical shaft drive plain bearings. The only parts that required replacing were the Oldham couplings which a toolmaker friend of mine had fitted to the shafts by spark erosion. Beats "manual stoning" as recommended by the manual!

I removed the oil pump from the bare crankcases by heating up in the oven after removing the four counter-sunk screws, then jolting it face down on the workbench. Works every time! The pump gears were in good condition, a little fettling with a Swiss file made for smooth running. I reassembled it with new brass shims, heated as for removal with two 3/16" dia. studs to aid location.

I had the magneto rewound and reconditioned by Ribblesdale Auto-electrics of Preston. Rebuilding the engine at this stage became a labour of love as I could see the end in sight.

My eldest son made me an engine stand, which was invaluable whilst setting the valves and ignition timing

The carburettor caused me some heartache even after buying all new parts for it. On firing up the bike for the first time, it would not respond to fine tuning and would not tick over at low revs. The only solution was either sleeving the carb or buying a new one. Amal do a fantastic service so, after much debate I bit the bullet and bought a brand new original spec carb. Good decision!

I bought a new wiring loom, light switches and control cables and had all the shiny bits re-chromed courtesy of Leigh Metal Polishers, Swinton, Manchester. I fitted a new rear tyre to match the new Avon front tyre. The bike was eventually fired up after some fine tuning with the ignition timing. I found out the hard way that you do not fit resister spark plugs and suppressor caps onto a magneto system.

The KSS has been shown at the Manchester Bike Show in 2017 & 2018 and actually ridden to local events around my home town. The climax was the Bike of the Year event in 2017.

The Velocette attracts a lot of attention where ever it goes, with minimum oil seepage, can't call it a leak, mainly from the kick start spindle (no oil seal in the early models). The paintwork, especially the petrol tank is of a very high standard, entrusted to RPM Racepaints of Leyland, whose skill in pin-striping and application of the Velocette stencil logo paint applied by airbrush to match the pinstripe, finished off with a resplendent coat of clear two pack varnish is remarkable.

This has been a journey of discovery, my last bike engine rebuild was a Triumph Thunderbird which was much simpler and cheaper than the Velocette, but not as much fun. Many thanks to the Velocette Owners Club and the members who have given me plenty of advice over the last few years. A good set of guys who are all true enthusiasts.

The VOC Spares scheme is also well appreciated, along with the spares specialists up and down the country who without their services, many restorations would be dead in the water.

Gordon Lund,
Preston, Lancashire.
Velocette Owners Club
Membership number: 3103

EXCEL London Classic Bike of the Year Competition 2017
One of top ten finalist voted by magazine readers

EXCEL Classic Bike of the Year
Commemorative Poster

Manchester Bike Show Event City2017
Class Winner for pre 1950 motorcycles

Being presented with winner notification from a nice young lady

and a trophy from the event organiser

My trophy

A local Vintage Vehicle Carnival 2017 at which the KSS took first prize against some stiff opposition

First prize in class at the Blackpool Motorcycle Show Stanley Park 2017

Blackpool Winter Gardens Motor Cycle Show 2017 Vintage Bike Display

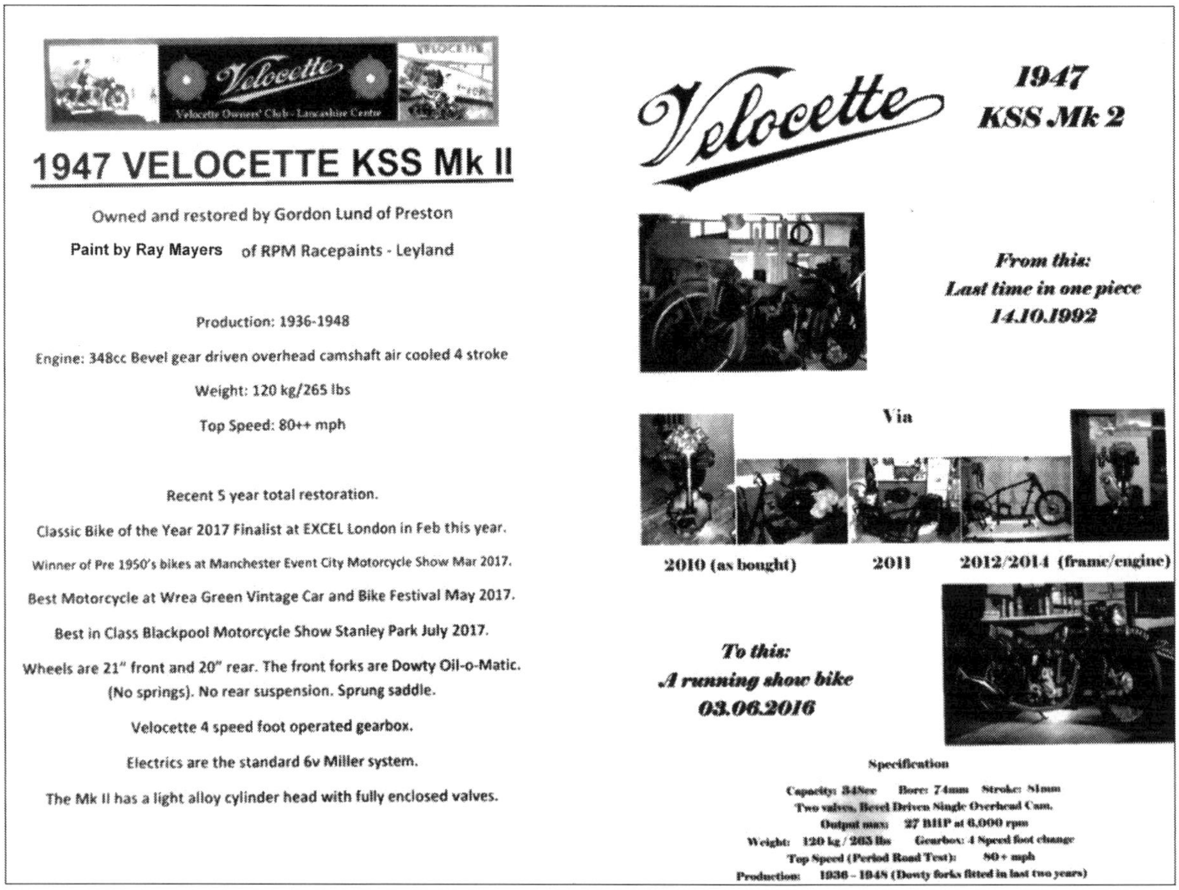

Velocette Owners' Club - Lancashire Centre

1947 VELOCETTE KSS Mk II

Owned and restored by Gordon Lund of Preston

Paint by Ray Mayers of RPM Racepaints - Leyland

Production: 1936-1948

Engine: 348cc Bevel gear driven overhead camshaft air cooled 4 stroke

Weight: 120 kg/265 lbs

Top Speed: 80++ mph

Recent 5 year total restoration.

Classic Bike of the Year 2017 Finalist at EXCEL London in Feb this year.

Winner of Pre 1950's bikes at Manchester Event City Motorcycle Show Mar 2017.

Best Motorcycle at Wrea Green Vintage Car and Bike Festival May 2017.

Best in Class Blackpool Motorcycle Show Stanley Park July 2017.

Wheels are 21" front and 20" rear. The front forks are Dowty Oil-o-Matic. (No springs). No rear suspension. Sprung saddle.

Velocette 4 speed foot operated gearbox.

Electrics are the standard 6v Miller system.

The Mk II has a light alloy cylinder head with fully enclosed valves.

Velocette

1947 KSS Mk 2

From this:
Last time in one piece
14.10.1992

Via

2010 (as bought) 2011 2012/2014 (frame/engine)

To this:
A running show bike
03.06.2016

Specification

Capacity: 348cc Bore: 74mm Stroke: 81mm
Two valves, Bevel Driven Single Overhead Cam.
Output max: 27 BHP at 6,000 rpm
Weight: 120 kg / 265 lbs Gearbox: 4 Speed foot change
Top Speed (Period Road Test): 80+ mph
Production: 1936 - 1948 (Dowty forks fitted in last two years)

Exhibition Display notice for public information

One Proud Owner

The Handlebar Controls

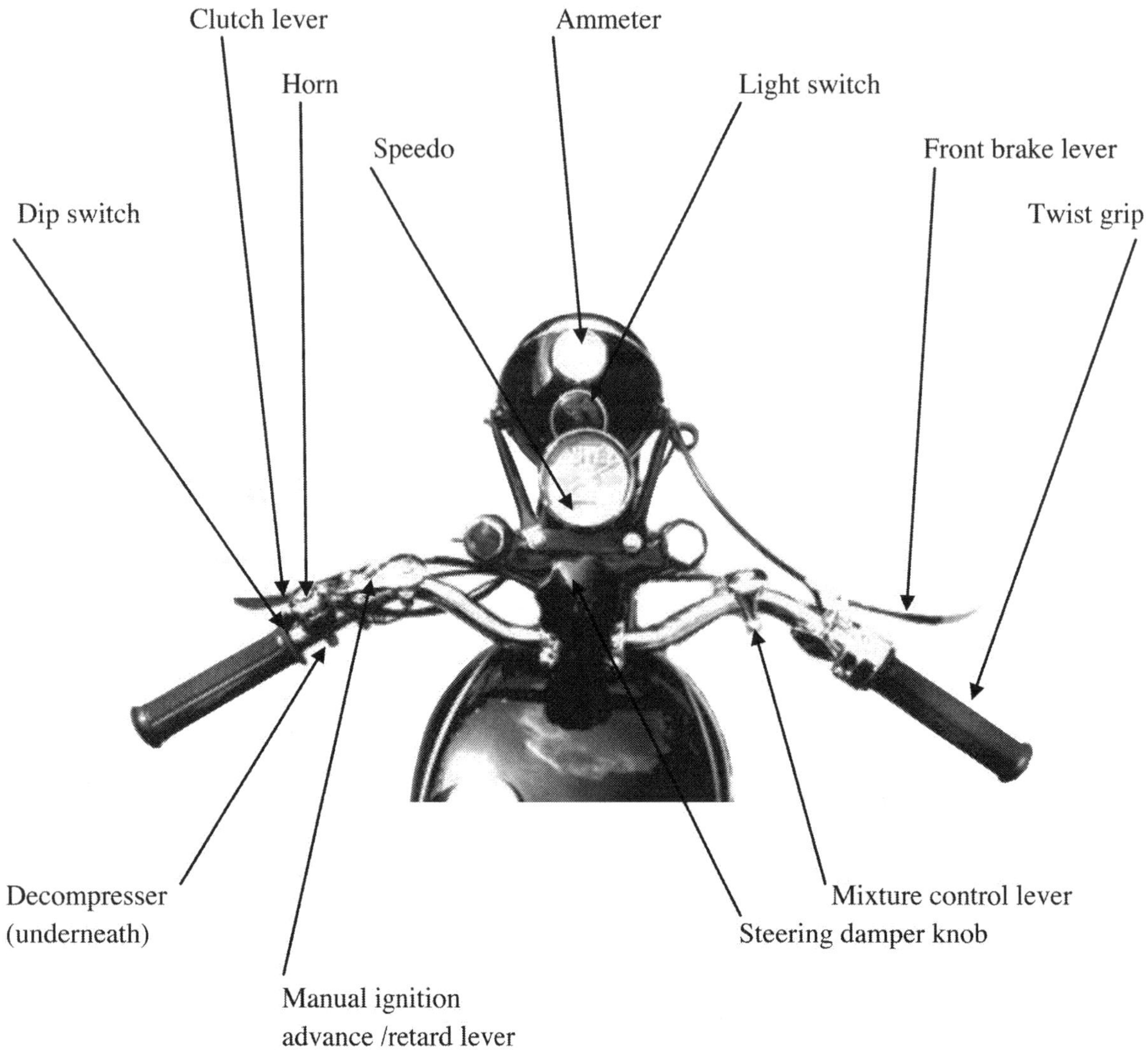

The Foot Controls

Gear lever

Rear brake lever

Further Velocette Valve Timing Comparisons

Cam No.	Inlet opens BTDC	Inlet closes ABDC	Exhaust opens BBDC	Exhaust closes ATDC	Clearance for checking timing		Running clearance		Ignition BTDC advanced	Model
					inlet	exhaust	inlet	exh		
K 17	37°	52°	65°	36°	.012"	.012"	.010"	.015"	42°	Early ohc (obsolete)
K 17/2	40°	56°	68°	47°	.012"	.012"	.010"	.015"	42°	ohc to end of 1931 (obsolete)
K 17/3	38°	60°	62°	40°	.012"	.012"	.012"	.020"	35°	DT engine (obsolete)
K 17/4	43°	70°	68°	48°	.012"	.012"	.012"	.022"	42°	KTT 1931
K 17/5	39°	69°	60°	40°	.012"	.020"	.012"	.020"	42°	KSS/KTS 1932-5
K 17/6	51°	57°	71°	43°	.020"	.020"	.015"	.025"	35°	KTT MkIV/MkV Type 30
K 17/7	35°	65°	70°	30°	.025"	.025"	.006"	.010"	30°	Replaced by K 17/10
K 17/8	55°	65°	75°	45°	.020"	.020"	.015"	.025"	32°	KTT MkVII/MkVIII
M 17	50°	60°	70°	40°	.010"	.015"	.003"	.006"	40°	MOV/MAC (rigid)
M 17/2	50°	60°	70°	40°	.015"	.020"	.003"	.006"	40°	MSS (Obsolete - use M 17/3)
M 17/3	30°	60°	60°	30°	.025"	.025"	.005"	.010"	40°	MSS (rigid)
K 17/9	30°	60°	60°	30°	.025"	.025"	.006"	.010"	40°	KSS/KTS (Obsolete)
K 17/10	34°	47°	64°	29°	.025"	.025"	.006"	.010"	30°	KSS/KTS MkII
K 17/11	55°	60°	75°	45°	.025"	.025"	.006"	.010"	30°	KTT MkVIII (latest type)
M 17/4	50°	60°	70°	40°	.030-.035"	.030-.035"	.005"	.005"	40°	MAC
M 17/5	30°	60°	60°	30°	.030"	.030"	.005"	.005"	40°	MAC alloy
M 17/6	55°	65°	75°	45°	.030"	.030"	.012"	.012"	36°	MAC (alloy - racing) Iron head
M 17/6	55°	65°	75°	45°	.030"	.030"	.006"	.006"	36°	MAC (alloy - racing) Alloy head
M 17/7	19°	49°	49°	19°	.030"	.030"	.005"	.005"	36°	MSS spring frame
M 17/8	55°	65°	75°	45°	.030"	.030"	.006"	.008"	38°	MSS Scrambler/VM/VR

NOTE: Ignition timing M 17/8 cam. Points to be open .0015" at 38° BTDC on full advance. Always ensure contact breaker gap is .012" before checking the timing. Take all slack out of gears when checking.

Note that my 1947 KSS Mk2 ignition timing advance is set at 38^0 BTDC which has since been found to be correct for that model (the figures in the table above are not to be used without expert verification)

THE SPORTY ALTERNATIVE

Velocette has always been a very race-oriented marque, aided in the 1920s by its fabulous KSS model.

The KSS was one of Velocette's earliest successes, and was born of a burgeoning interest in the sporty overhead-camshaft (ohc) machine that the company developed quite a reputation for making in the mid to late 1920s, and just beyond.

During 1925, interest was mounting in the British bike industry about a new 350cc single-cylinder Velo. In June of that year the Model K emerged, featuring a plucky ohc engine designed by Percy Goodman. The engine in particular was viewed as a great achievement on his part. The Model K formed the blueprint for all the K models that would appear in the next 25 years, but the KSS itself arrived very shortly after the original Model K.

In its infancy the KSS was often referred to as the MK1 - the 'Super Sports' SS suffix would take root properly in the

Availability ●●●●●	Pricing	High: 1928 Mk1 £15,000
Desirability ●●●●●		Low: 1948 Mk2 £12,000

years that followed. In 1926, a brand new KSS straight off the production line at Hall Green would have cost you £75.

Its engine was basically the same Percy Goodman-designed creation that powered the Model K, though the KSS engine had a higher compression ratio and slightly modified camshafts. In terms of appearance it differed from the Model K in that it had narrower mudguards and no rear carrier, making it slightly lighter and a touch sportier. It also came with an 80mph guarantee, which would have been rather remarkable had the Brough Superior SS100 and its 100mph guarantee not arrived during the same year.

The various K models generated a lot of interest due to their potential as racing machines, and after Alec Bennett's TT win in 1926 - a race he won on an ohc Velocette - they became even more popular as racers. It was this TT win that heralded the following years of sporty Velocette motorcycles.

The early KSS machines were outfitted with a three-speed gearbox, but in later years the gearbox grew to a four-speed assembly. The two and a quarter gallon tank that had arrived in 1928 was replaced by a much larger four gallon tank in 1935. These amendments were only two of the many changes made by Velocette that resulted in the machine becoming heavier and, ultimately, much slower. The KTT was running rings round the KSS as early as 1928, before it went on to be Velocette's flagship model for several years. During this period, the KSS relaxed into more of a sports tourer role, as it could no longer really compete with its brother as a thoroughbred racing machine.

Velocette attempted to inject some vigour into the KSS in 1935 and debuted the Mk2, which had a completely new alloy cylinder head, fully enclosed valve gear and downdraught inlet track. In this form, the KSS continued until the outbreak of the Second World War, then production restarted after the cessation of hostilities, with only minor modifications.

Velocette managed to manufacture roughly 6000 KSS machines at Hall Green, and both 'marks' are highly sought-after today.

Extracts from a booklet titled; Memorable British Bikes courtesy of The Classic Motorcycle

The KSS Mk 2 Engine

My KSS Mk2 engine on the bench in its engine stand

Old Velocette Adverts

Velocettes were very colourful in their advertising department be it that most of the bikes were finished in black enamel paint with gold pin striping and lettering. It was not until much later in the swinging sixties that brighter colours were available. Red, blues and greens abounded but by then it was all over and the far eastern makes took over.

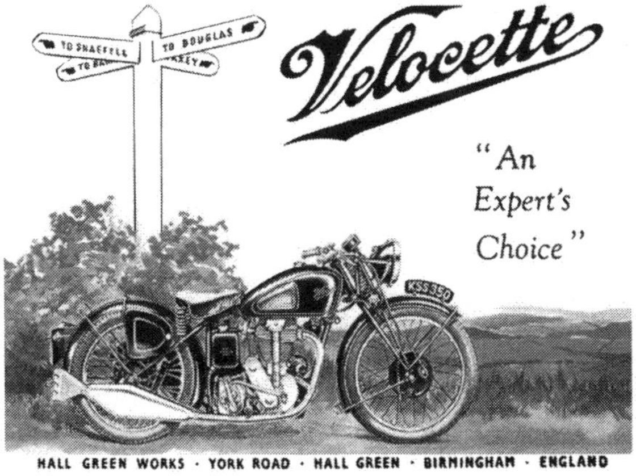

Competition Velocettes
The Anatomy of a Velocette KTT MkVIII race bike

Geoff Bloor's immaculate 1947 KTT Mk VIII racing motorcycle
Winner of the best competition bike at the 2018 Event City Manchester Show

Keith Riley's race prepared KSS Mk 2 A good contender in Classic Race events.

A Mk VIII KTT on the track, where it belongs!

The Ultimate Velocette Race Bike "The Roarer".
The 1950's attempt at 500cc GP. Supercharged which was subsequently banned.
Normally aspirated could not match the Italian 4 cylinder machines.

The KTT Mk VIII

The following pages are a unique presentation of the ultimate 350cc single cylinder over head camshaft motorcycle engine, available to the public at that time, as an over the counter, ready to race and win bike.

In its heyday it won many races and championships and was only out shone in later years when the Velocette race team just could not compete with the likes of Norton and the Italians who had far bigger budgets.

The KTT and its road going brother the KSS parted company a long time ago but very often shared common components throughout their development. At one stage, the KTT MkVI shared the same cylinder head with the KSS Mk 2 whilst the factory race team were developing their own cylinder head for the next KTT model notable for the large fins to provide extra cooling under race conditions.

The KTT Mk VI. Note the KSS Mk 2 Cylinder head

Bike on display at the National Motorcycle Museum

The 348 c.c. Overhead camshaft
KTT VELOCETTE ...
The Whys and Wherefores of a Famous Engine Designed Specifically for Road Racing
by "UBIQUE"*

Everyone knows that, for years, the KTT Velocette was unique in being the nearest possible thing to a genuine T.T. model which was available to the public. It was, in fact, designed and intended for road racing, and this must be remembered when considering the following notes.

Looking at all the parts neatly laid out on the sheets of brown paper, I was at a loss to know where to begin. Finally, after discussing the point with Messrs. Percy Goodman and Harold Willis-now, unhappily the late Harold Willis - I decided to start at the top and work down, yet in actual practice my first question dealt with the cylinder barrel. "What are the materials, and how are they mated?"

Rather to my surprise I learned that the silicon-aluminium-alloy jacket with its very deep radiating ribs was "cast on" the nickel cast-iron barrel, with its corrugated outer wall, being heated and placed in the mould before the alloy was poured.

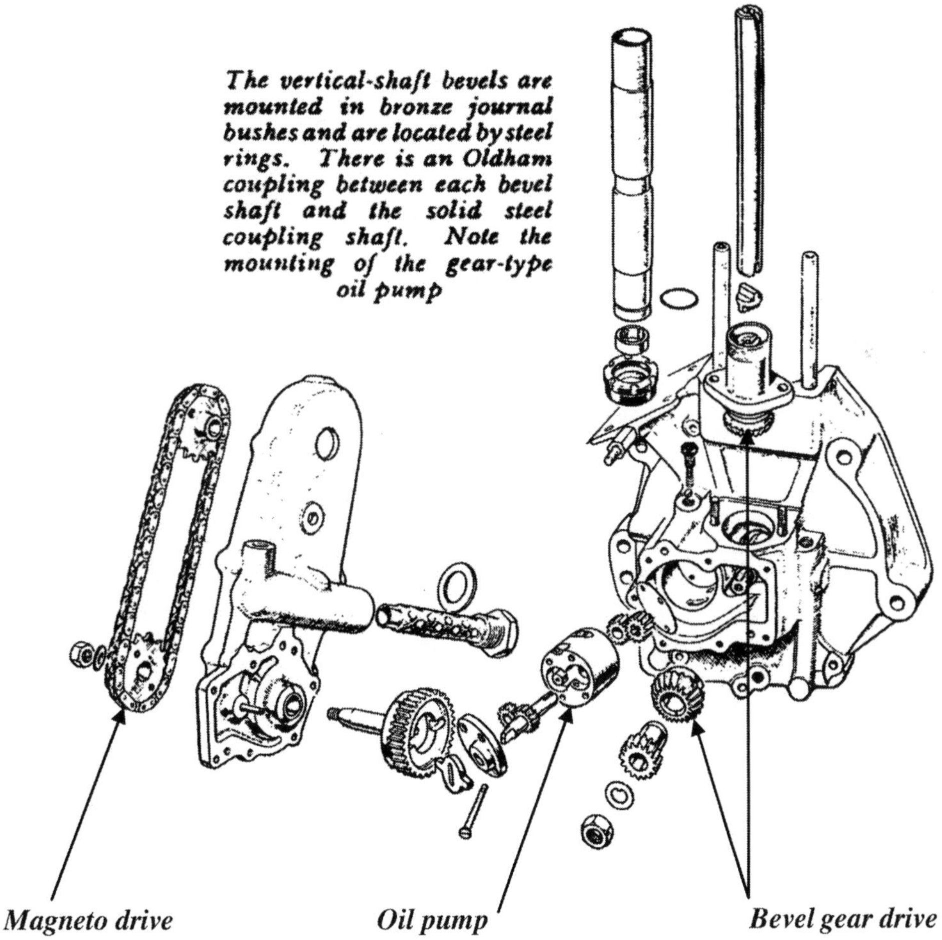

The timing side crankcase of the KTT engine showing the bevel geared drive to the O.H.C. valve gear, oil pump and magneto chain drive

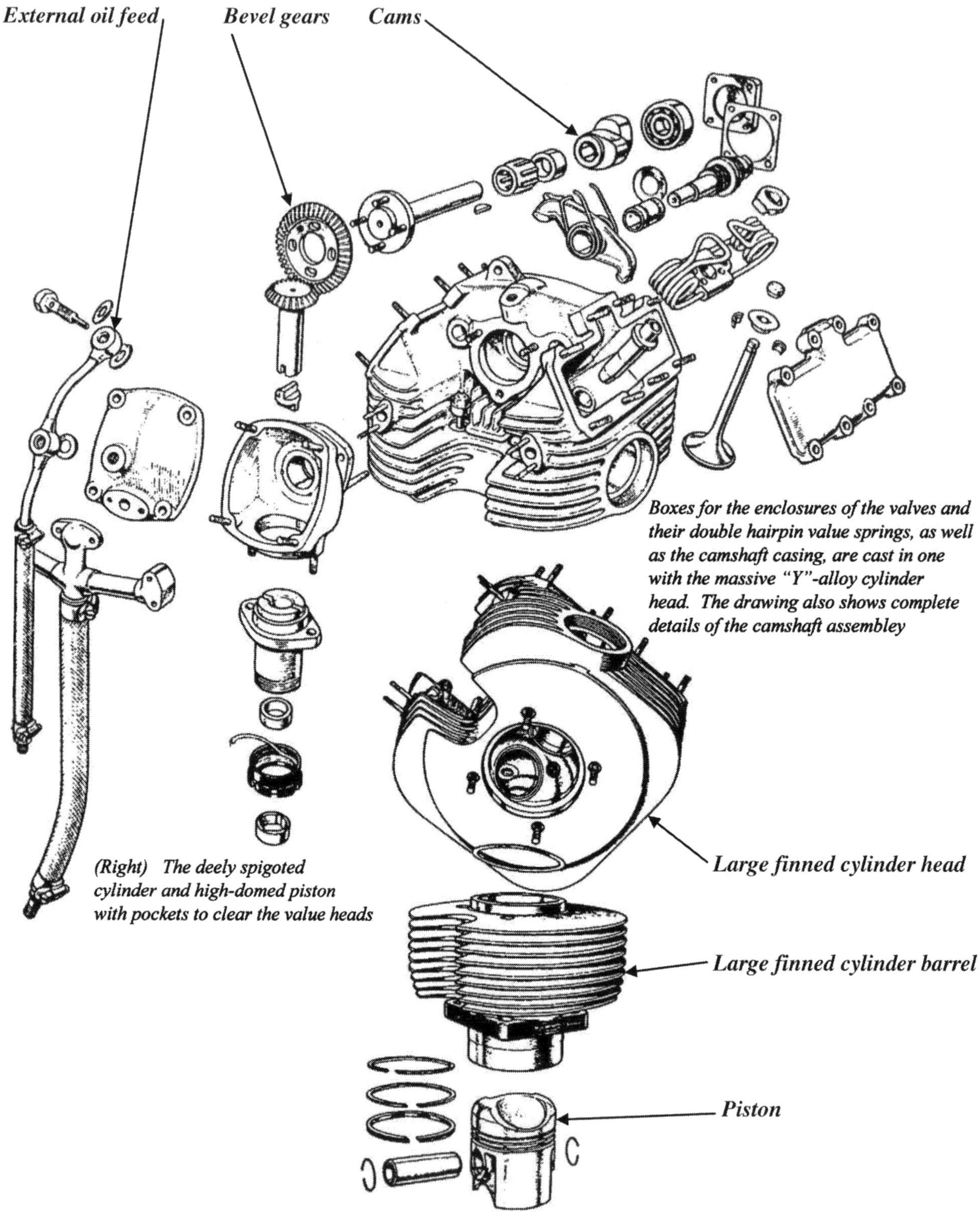

Boxes for the enclosures of the valves and their double hairpin value springs, as well as the camshaft casing, are cast in one with the massive "Y"-alloy cylinder head. The drawing also shows complete details of the camshaft assembley

(Right) The deely spigoted cylinder and high-domed piston with pockets to clear the value heads

The cylinder, piston and cylinder head complete with valve gear.
The KTT engine notable differences with the KSS engine are the large fins on the aluminium barrel and cylinder head along with external oil feeds to the top bevel gears and camshaft

Mr. Willis did not think there was much to choose between this practice and that of pressing-in a liner or shrinking-on a jacket, but actually Velocettes, he said, had had slightly better results with the practice outlined; it prevents the possibility of the liner "creeping."

The head, I gathered, was of heat-treated "Y"-alloy, a metal chosen for its light-weight, strength and high thermal conductivity. The head ribs of the KTT are roughly 9 inch square in plan view, and the camshaft casing and the boxes for the enclosure of both valves and their double hairpin valve springs are cast with the head, so you can judge the importance of a lightweight material.

Of course, I asked why these huge cooling fins were employed, suggesting that it must be almost impossible for a considerable draught of air to reach the rib roots.

In this matter Mr. Willis told me I was correct, but that careful tests showed that it was of greater importance to get a big cooling surface right out into the draught than to rely on such air as might possibly pass the front wheel, mudguard, fork, frame, etc., reaching the roots of the fins. That explains the importance of high conductivity, does it not?

Again, he pointed out that the joint between the head and barrel, at the top of the spigot, consisted of a laminated copper washer. Each of the four laminations is about 0.009 inch thick, and this arrangement makes a very satisfactory job with direct metal-to-metal contacts throughout.

Internally, the head is approximately hemispherical, the long-reach 14mm plug being placed on high up and screwed directly into the head.

My next question was, "Why do you employ different materials for exhaust and inlet valve seating?"

"A hard aluminium-bronze alloy is used for the exhaust seat, because of the high conductivity of this metal helps to keep the exhaust valve cool and, incidentally, its high co-efficient of expansion helps to maintain good contact with the head casting at high temperatures. For the inlet valve, the seat of which is apt to be damaged by grit, a nickel cast-iron is employed on account of its hardness."

Pressed-in Seatings

"How are the seatings help in position?" I then asked.

"The head is heated in an electrical furnace to a temperature of 200 degrees C., and the seatings previously mounted on special jigs, are pushed home very quickly. Speed is essential in this process, for as soon as the seatings come in contact with the hot head, they also expand and lock in position."

The pale gold of aluminium-bronze valve guides next caught my eye. The high heat conductivity and high coefficient of expansion, plus the fact that this material is an excellent bearing metal, make it very suitable for the purpose. Add to this, ample lubrication from the cam-box and the very high finish of the valve stems and guides, and I was not surprised to hear that wear at this point is negligible over long periods.

Incidentally, the valve stems work with quite close clearance limits, the actual figures being 0.0015in for the inlet and 0.003in for the exhaust.

"Why is the inlet valve of larger diameter that the exhaust?" Both my mentors answered this question. Perhaps I may summarize their remarks by saying that it is far more difficult to fill a cylinder by atmospheric pressure than to empty it when the piston pushes behind and there is help at certain vital speeds from a specially designed exhaust pipe.

"The valves themselves? Well, the inlet is made of a cobalt-chrome steel having high tensile strength and hard wearing surfaces. It has a throat diameter of $1^9/_{16}$ inch, and the stem is reduced to the smallest possible diameter consistent with safety, so as to reduce port obstruction. K.E.965 steel was chosen for the exhaust valve because it retains great strength at the very high temperatures at which it is called upon to work.

Valve Timing

The port diameter of the exhaust valve is $1^7/_{16}$ inch, and the stem is approximately $^1/_{16}$ inch larger than that of the inlet, and the lift of both valves is $^3/_8$ inch. Both have moderately recessed heads, but whereas the end of the inlet stem is hardened, the material of the exhaust is unsuited for this purpose, so a hardened end-cap is used.

I asked if the valve timing were a secret, but it was given to me without the slightest hesitation. Checked with a tappet clearance of 0.020inch in each case, the timing is as follows: inlet opens 55° BTDC, and closes 65° ABDC.; exhaust opens 75° BBDC. & closes 45° ATDC.

One hundred degrees overlap. Think of it! That ought to get the inlet gas column moving and ensure that the valves are well open when they will do most good!

The correct running tappet clearances are: inlet 0.015in, exhaust 0.025in.

Even the details of the valve spring cap and collect are of unusual interest. It was pointed out that the groove in the top of the valve stem is extremely shallow in order to reduce the strength of the stem as little as possible. To make up for this, the split collett is so designed as to grip the stem, and under the wedging action of the spring cap, this grip is almost sufficient without the groove.

Free to Rotate

The item which corresponds to the usual spring cup (and is described as such above) does not, in fact, form a direct thrust face for the hairpin springs, the looped ends of which are hooked on to a separate steel spring abutment, loosely mounted round the "spring cup." Thus, at the expense of a little extra reciprocating weight, the valve is free to rotate, a most desirable feature.

At their lower ends the valve springs are tucked into longitudinal holes drilled in a steel plate which surrounds the upper part of the valve guides and these plates have at each end, a tapped hole to carry the fulcrum of a spring removing tool.

"What spring pressures are necessary for this engine?" I asked.

"Measured with the valve closed, a pressure of 110lb, plus another 25lb for the return spring on the rocker, the latter is an important point which we will discuss when we get there."

"Does the total enclosure of the valve gear cause any kind of trouble in the way of overheating?"

"No, the valve boxes have a large external area exposed to the draught and the interior is cooled by a liberal circulation of oil."

"Some years ago you used a two-cam-shaft racing engine," I said. "Why, in this engine, have you reverted to the single-camshaft type?"

"I was about to mention that very point," said Mr. Percy, "as it is a matter of some historical interest. The twin-camshaft job, in spite of its theoretical advantages, presented certain practical difficulties. (Since overcome, as recent Velocette racing engines have demonstrated.-Ed) It was difficult to provide reasonably accessible tappet adjustment, and to enclose the mechanism-a point which, as you know, we consider to be important. Further, for the first time we had trouble with the vertical shaft drive, which could be traced indirectly to the upper works of the valve gear."

Mr. Willis continued: "After much experiment we found that the best of all round results could be obtained from a single-camshaft engine if the rockers could be held in constant contact with the cam. This led to the powerful hairpin rocker return spring which you are examining. At first this caused scoring and rapid wear of the rocker heels and cams, and we found it necessary not only to provide a jet of oil directly on to the point of contact between cam and rocker, but also to face the heel of the rocker with Stellite. These precautions have been completely successful.

Floating Bushes

On inquiry, I found that these rockers are made from stampings of air-hardening, nickel-chrome steel of 100-ton tensile test. Between each rocker and its massive eccentrically mounted pivot-pin is a bronze floating bush.

"Why are the cams made separate from their shaft?"

"Well, the cams, which are designed to give constant valve acceleration and deceleration are of a straight carbon-steel, deeply cased, and have 1 per cent carbon in the case. The shaft is of 3 per cent nickel steel, case-hardened so as to form a track for the driving-side roller bearing. The cams are pressed and keyed on to the shaft. The rollers on the drive side are caged and a ball-bearing on the other end of the shaft takes the end thrust from the bevel gears."

"What about the vertical-shaft drive?"

"Most of it is standard, including the upper pair of bevels. The crown bevel is bolted to its back plate by four bolts, and the bolt holes are elongated to provide a half tooth adjustment for timing. The lower bevels have been strengthened to withstand the extra strains imposed by high speed and heavy valve springs."

"Why the lower bevels and not the upper ones?"

"Because the shape of the small diameter bevels of approximately equal size is of not so favourable to tooth strength."

"I note you retain the hunting tooth."

"Yes, the odd teeth in the vertical shaft drive distribute the loading as evenly as possible. The vertical-shaft bevels are mounted in bronze journal bushes, the thrust being taken on the end faces. Steel rings are lightly pressed on to the shafts to locate the bevels length-wise in their bearings. There is an Oldham coupling of oil-hardened nickel-chrome steel between each bevel shaft and the solid steel coupling shaft. This coupling shaft is mounted in plain bearings in its enclosing tube, and the tube is held at each end by asbestos-packed gland-nuts, and although it appears to be quite rigid, there is just a trace of intentional flexibility in the mounting."

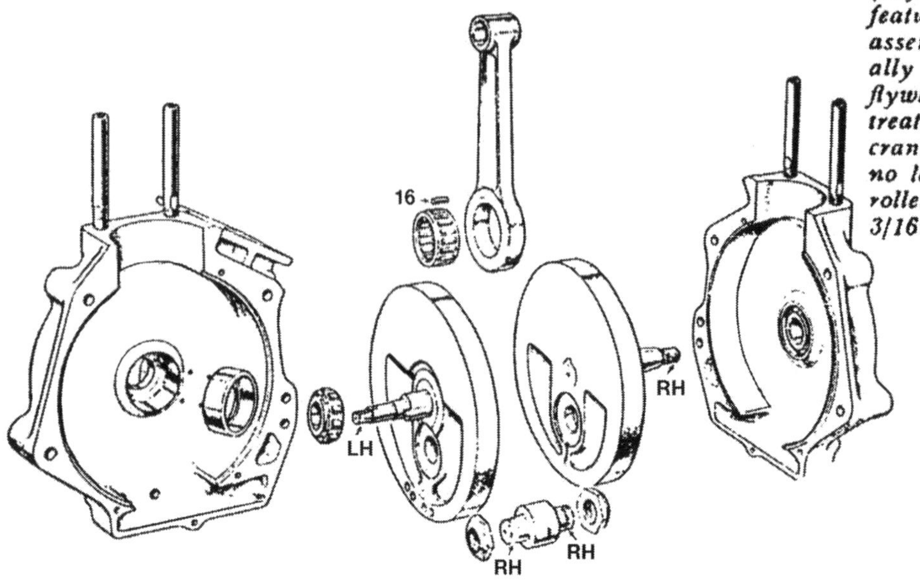

(Left) Immense rigidity is a feature of the crankshaft assembly and of the unusually narrow crankcase. The flywheels are made of heat-treated carbon steel. while the crankpin has a diameter of no less than 1⅜in. Big-end rollers measuring 9/16in x 3/16in are located in a Duralumin cage

KTT engine crankshaft assembly

The unusually narrow aluminium-alloy crankcase is noticeable without need of questions, and the single-row caged roller bearings on each side are supported directly under the massive walls in which the main strength of the crankcase lies. The only additional stiffening is provided by shallow webs on the outside of the drive side. As in other Velocette models, the primary drive lies inside the final drive, and therefore the engine sprocket is close up to the main bearing. Thus, there is very little overhang and no need for a multiplicity of crank bearings.

Immense Rigidity

A glance at the compact and immensely rigid crank unit with its single-row big-end and comparatively narrow (13/16in) flywheels, also at the smooth, web-less interior of the crankcase, shows how that effect of slimness is attained, but here more questions were necessary.

"What are the flywheels made of?"

"Heat-treated carbon steel."

"Why heat-treated?"

"To prevent stretch in the shaft holes."

"How are the crank axles fixed in the flywheels?"

"Just pressed in, and located by a pin."

"The pin is really a screw lying parallel with the shaft, partly in the shaft and partly in the flywheel boss. These shafts, of 3 per cent nickel steel, are very slightly tapered so that the inner races of the main bearings are securely locked when pushed home. The outer races are shrunk in to the crankcase, so that they also are firmly held."

Then I asked about the massive crank-pin, which adds so materially to the rigidity of the crank unit. It is made of nickel-chrome steel, case hardened to form the inner race of the big-end bearing. The main diameter is $1^3/_8$ inch, and the ends are very slightly tapered and pulled into the flywheels against wide shoulders. The nuts which fix it are of heat-treated nickel steel.

Rollers of $^9/_{16}$ inch x $^3/_{16}$ inch are employed in the big-end bearing. These rollers are located in a Duralumin cage, the bars of the cage being relieved on their inner surface so that only the end rings of the cage bear on the crankpin.

The reason for this is that the Duralumin cage is apt to wear the hardened crankpin, and the relief of the bars prevents damage to the roller track. This, of course, is a fact, though it may seem hard to believe that more wear is caused by the light-alloy cage than the heavily loaded rollers. The outer race of the bearing is pressed into the connecting rod big-end and a bronze bush into the small-end.

The massive forged connecting rod is made from an oil hardening nickel-chrome steel, heat-treated to about 80 tons tensile. "Of course," said Mr. Willis, "we could use a steel with higher tensile strength, but tensile strength is not the only necessity, for great toughness is also required. The rod is machined all over and polished on the outside. This is not only useful for removing surface scale and weight, but also tends to reveal surface cracks or flaws."

Of case-hardened nickel-chrome steel, the gudgeon pin is no less than $^{13}/_{16}$ inch diameter; it is hollow and is taper bored at the ends to reduce weight. It floats in the little-end of the connecting rod, and is retained in the piston bosses by spring-wire circlips. As was pointed out, the outer extremities of the gudgeon pin are so chamfered that any tendency for the pin to move endways tends also to jam the circlips in their grooves, rather than to displace them. In order to provide the standard compression ratio of approximately 10.9 to 1, the piston head must be steeply domed. It is! So much so, indeed, that circular pockets are formed in the sides of the dome to clear the valve heads.

The skirt is of slipper form, and the whole is made of heat treated "Y"-alloy, sand-cast,

"Why sand-cast?" I asked.

"Because it has given us better results that way," was the reply.

There are small, vertical ribs from the gudgeon-pin bosses upwards, and there is a circular rib at the level of the boss centres for maintaining the shape of the slippers. Two narrow pressure rings of Wellworthy "Themocrom" (heat-formed, not hammered) and a slotted scraper ring are fitted.

"Each engine, individually, is adjusted by shims under the cylinder base so that the cylinder head capacity is 35c.c.," said Mr. Willis.

This 348c.c. engine develops 27 b.h.p. at 6,500 r.p.m. It may be run at 7,000 r.p.m. in the normal course of events, and may be run with safety up to 7,500 r.p.m. for short periods, and will run up to 8,000, though this is considered to be beyond the limits of reasonable safety.

A diagrammatic view of the lubrication system of the KTT engine, showing the feeds to the big-end bearing and to the overhead camshaft and rocker gear. The oil is pumped at the rate of 1/6th gallon a minute to a check valve, the object of which is to prevent oil seepage from the tank to the crankcase when the engine is stationary

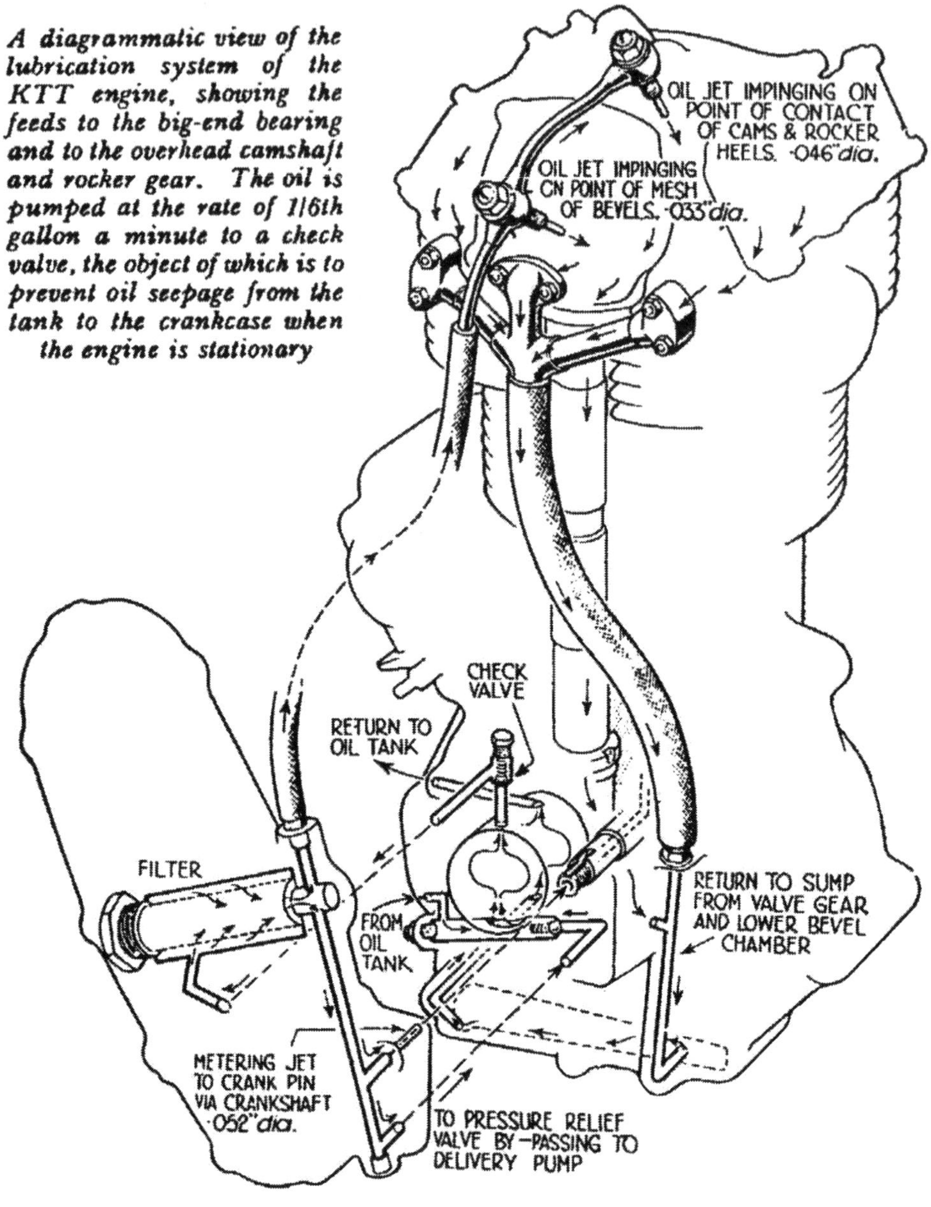

The KTT engine oil lubrication system. Designed with high speed running in mind

****By "Ubique", a nome-de-plume by a journalist for "The Motor Cycle" His real name was Henry Teage and was at one time the Technical Editor of "The Motor Cycle"***

Taken from "Motor Cycle Engines", published by The Motor Cycle, in 1951.
Acknowledgement to Mortons Motorcycle Media, holders of the copyright.

Modern Wiring Diagram (Solid State Voltage Regulator)

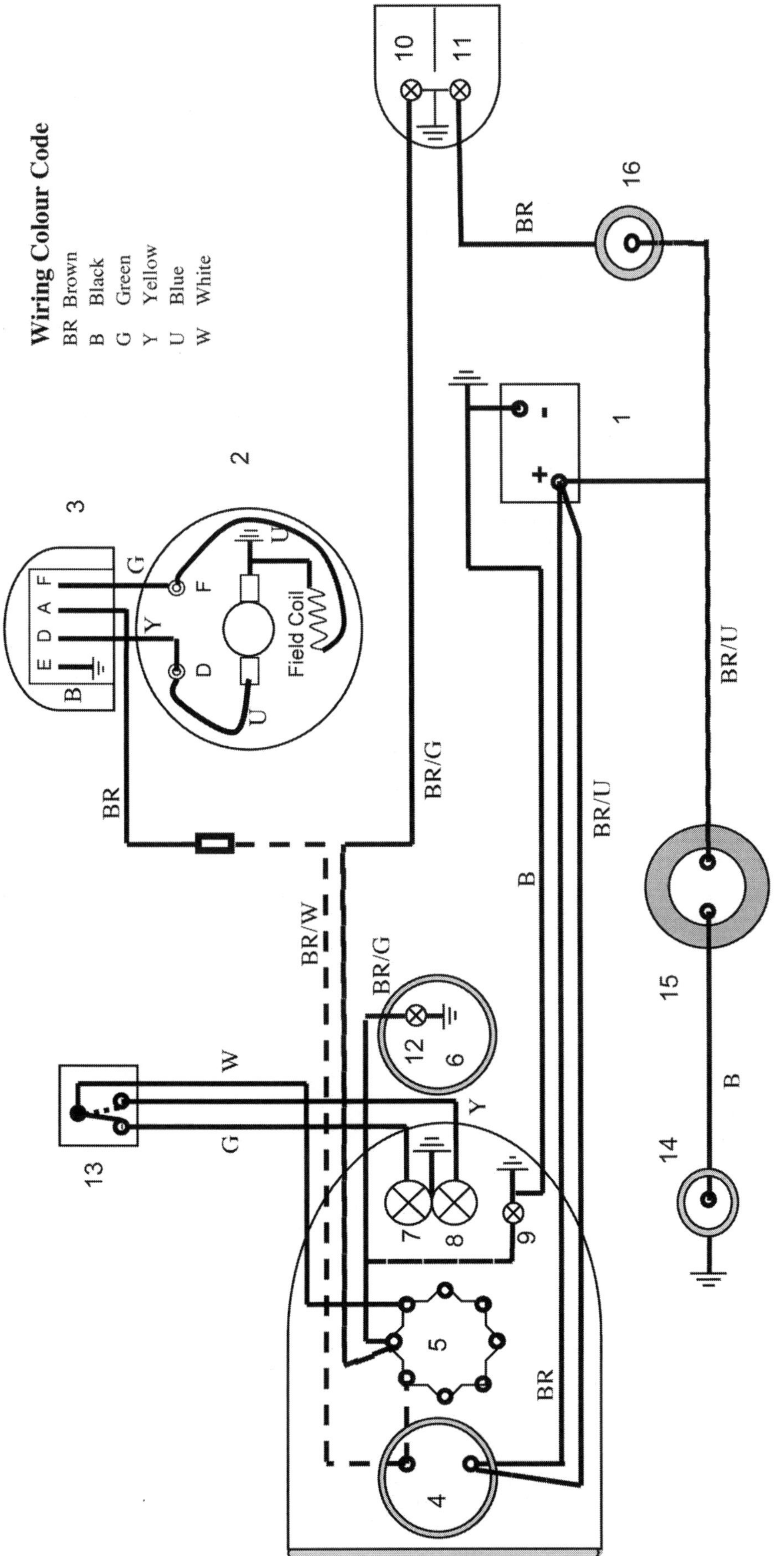

1 - Battery
2 - Dynamo
3 - Solid State Voltage Regulator
4 - Ammeter

5 - Light Switch
6 - Speedometer
7 - Headlight bulb dip
8 - Headlight bulb main beam

9 - Pilot bulb
10 - Tail light bulb
11 - Stop light bulb
12 - Speedometer bulb

13 - Dipswitch
14 - Horn push button
15 - Horn
16 - Stop light switch

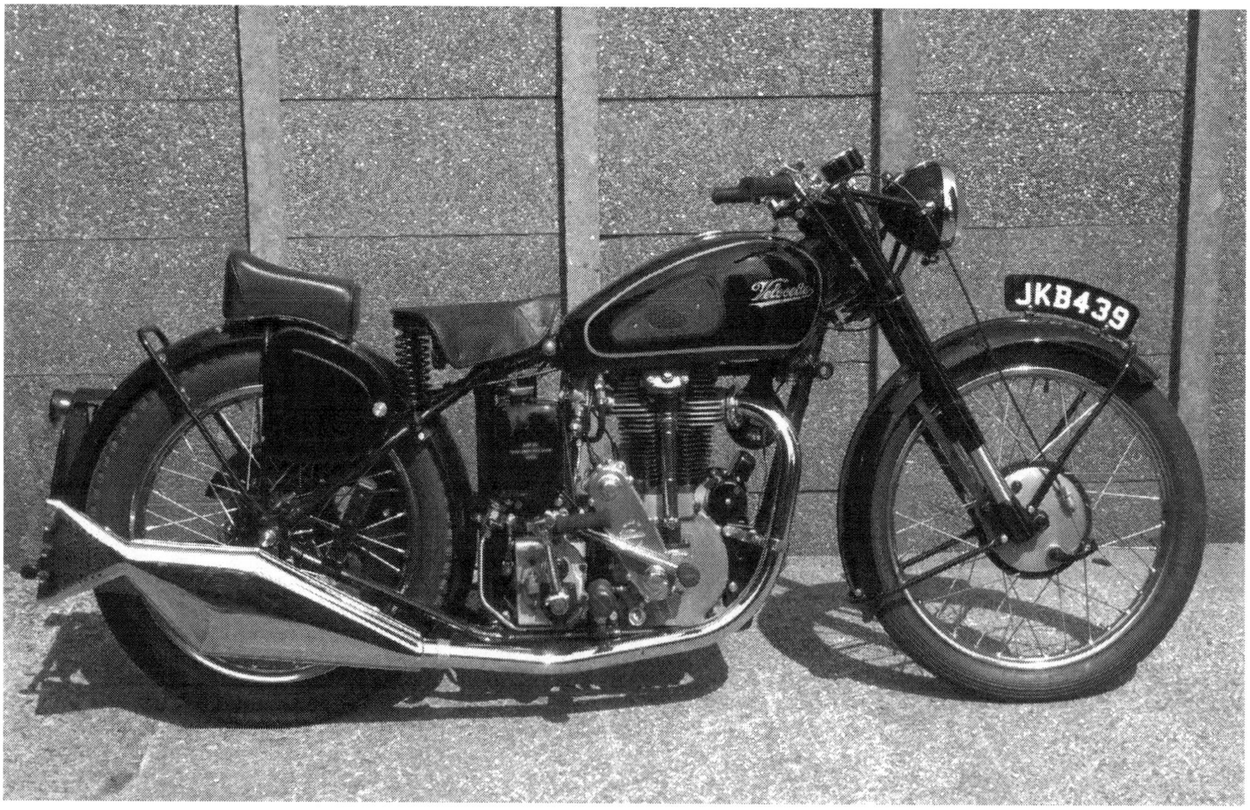

My 1947 KSS Mk2 parked in my drive ready for another run out.

Made in the USA
Monee, IL
05 November 2026

67815697R00079